SpringerBriefs in Applied Sciences and Technology

Computational Intelligence

Series Editor

Janusz Kacprzyk, Systems Research Institute, Polish Academy of Sciences,
Warsaw, Poland

SpringerBriefs in Computational Intelligence are a series of slim high-quality publications encompassing the entire spectrum of Computational Intelligence. Featuring compact volumes of 50 to 125 pages (approximately 20,000-45,000 words), Briefs are shorter than a conventional book but longer than a journal article. Thus Briefs serve as timely, concise tools for students, researchers, and professionals.

Oscar Castillo · Patricia Melin

Type-3 Fuzzy Logic in Time Series Prediction

 Springer

Oscar Castillo
Division of Graduate Studies
Tijuana Institute of Technology, TecNM
Tijuana, Baja California, Mexico

Patricia Melin
Division of Graduate Studies and Research
Tijuana Institute of Technology, TecNM
Tijuana, Baja California, Mexico

ISSN 2191-530X   ISSN 2191-5318 (electronic)
SpringerBriefs in Applied Sciences and Technology
ISSN 2625-3704   ISSN 2625-3712 (electronic)
SpringerBriefs in Computational Intelligence
ISBN 978-3-031-59713-8   ISBN 978-3-031-59714-5 (eBook)
https://doi.org/10.1007/978-3-031-59714-5

This Springer imprint is published by the registered company Springer Nature Switzerland AG
The registered company address is: Gewerbestrasse 11, 6330 Cham, Switzerland

Paper in this product is recyclable.

Preface

This book focuses on the field of type-3 fuzzy logic for applications in time series prediction. The main idea is that a higher type and order of fuzzy logic can help in solving various prediction problems and find better results. In addition, neural networks and fractal theory are employed in enhancing prediction results. In this regard, several hybrid intelligent methods are offered. In this book we test the proposed methods using several prediction problems, like predicting COVID-19 and the stock market. We can notice that when Type-3 fuzzy systems are implemented to model the behavior of systems, the results in prediction are enhanced, because the management of uncertainty is better. For this reason, we consider in this book the proposed methods using type-3 fuzzy systems, neural networks and fractal theory to improve the prediction behavior of the complex nonlinear systems.

This book is intended to be a reference for scientists and engineers interested in applying type-3 fuzzy logic techniques for solving complex prediction problems. This book can also be used as a reference for graduate courses like the following: soft computing, fuzzy logic, neural networks, bio-inspired algorithms, intelligent prediction, and similar ones. We consider that this book can also be used to get novel ideas for new lines of research, or to continue the lines of research proposed by the authors of the book.

In Chap. 1, we start by offering a brief introduction of the potential use of Type-3 Fuzzy Systems for complex problems in time series prediction. We also mention other possible applications of the proposed prediction approach.

We describe in Chap. 2 the basic concepts, notation, and theory of type-3 fuzzy logic, and fuzzy prediction. This chapter reviews the background, main definitions and concepts, useful for the development of this research work.

We describe in Chap. 3 a review of the existing works on type-3 fuzzy theory applied to time series prediction in the current state of the art. In addition, an analysis of the current and future trends in this area are mentioned.

Chapter 4 is devoted to describe the proposed approach for employing a Hybrid Interval Type-3 Fuzzy-Fractal Approach in prediction. In this case, type-3 fuzzy logic and the fractal dimension are combined for achieving accurate prediction. In particular, the prediction of COVID-19 is utilized to illustrate the proposed approach.

We offer in Chap. 5 an approach for type-3 fuzzy aggregation of neural network predictors. Type-3 fuzzy aggregation of neural network was tested with the COVID-19 and Dow Jones time series and the results surpass other methods.

We explain in Chap. 6 an approach with type-3 fuzzy aggregators for neural network ensembles in Prediction. Additionally, the type-3 fuzzy approach for neural ensembles was applied in predicting the COVID-19 time series with good results.

We describe in Chap. 7 an approach for optimal type-3 fuzzy systems and ensembles of neural networks using the firefly algorithm and the COVID-19 information was employed to perform times series prediction. As the structure of the neural networks and fuzzy system were optimized, the prediction results for many countries outperform previous methods.

We describe in Chap. 8 the conclusions of this work on type-3 fuzzy prediction, as well as some future research works are envisioned.

We end this preface of the book by giving, thanks to all the people who have help or encourage us during the writing of this book. First of all, we would like to thank our colleague and friend Prof. Juan Ramón Castro for always supporting our work, and for motivating us to write our research work. We would also like to thank our colleagues working in Soft Computing, as they are more in number to mention. Of course, we need to thank our supporting agencies, CONACYT and TecNM, in our country for their help during this project. We have to thank our institution, Tijuana Institute of Technology, for always supporting our projects. Finally, we thank our respective families for their continuous support during the time that we spent in this project.

Tijuana, Mexico   Prof. Oscar Castillo
February 2024   Prof. Patricia Melin

Contents

Chapter 1
Type-3 Fuzzy Prediction

The essential constructs in type-3 fuzzy logic and their utilization in prediction are offered in this monograph. The focus is on the fundamental reasons for utilizing type-3 in achieving an accurate prediction. Type-3 is a novel theory to model uncertainty that can be utilized in prediction. Type-2 have been previously used as a way for considering prediction, but recently type-3 offers an alternative in considering more complex prediction problems. In this work, we review the constructs of type-3, which are studied in a more thorough way. In addition, we formulate a way for constructing type-3 systems and highlight them with prediction and forecasting problems.

Uncertainty has effect on decision-making and manifests in many ways. The concept of information is inherently associated with the concept of uncertainty [1, 2]. Uncertainty is an attribute of information [3]. For systems based on type-1 sets, an uncertainty model with values in [0, 1] is utilized. When an entity is uncertain, like a measurement, it is difficult to specify its exact value, and of course a type-1 fuzzy set makes more sense than a traditional set [3, 4]. However, it is not reasonable to use an accurate membership function for something uncertain, so in this case what we need is another type of fuzzy sets, like those which are able to handle these uncertainties, and the so-called type-2 fuzzy sets [5, 6] were put forward for achieving this purpose. The amount of uncertainty in a system can be reduced by using type-2 fuzzy logic because this logic offers better capabilities to handle linguistic uncertainties by modeling vagueness and unreliability of information [7, 8]. Recently, there has been some works dealing with interval type-3 fuzzy models, as they can provide even better capabilities for handling uncertainty in prediction applications [9] and other areas, and this is the main motivation of this book.

Type-2 fuzzy models have emerged as an interesting generalization of fuzzy models based upon type-1 fuzzy sets [5, 10]. There have been a number of claims put forward as to the relevance of type-2 fuzzy sets being regarded as generic building constructs of fuzzy models [11–13]. Likewise, there is a record of some experimental evidence showing some improvements in terms of accuracy of fuzzy models

O. Castillo and P. Melin, *Type-3 Fuzzy Logic in Time Series Prediction*,
SpringerBriefs in Computational Intelligence,
https://doi.org/10.1007/978-3-031-59714-5_1

of type-2 over their type-1 counterparts [14–18]. There have been a lot of applications of type-2 in intelligent control [19–24], pattern recognition [25–29], intelligent manufacturing [8, 16, 30], time series prediction [14, 31], and others [8]. Recently, there has been some evidence [9] that type-3 fuzzy systems can improve results with respect to type-2 in some cases, and for this reason the importance of this book that provides the basic concepts to develop these type-3 fuzzy systems for prediction.

Methods for building type-2 models are put into two classes. The first one starts from an initial type-1 model and subsequently a type-2 model is obtained by extension. In the second case, the model is elucidated directly from data. In particular, metaheuristics could be employed in searching for the appropriate model. Now for type-3, we start with an initial type-2 model to get to type-3. We are considering several prediction applications in this book, but we plan to consider in the future more complex prediction problems, such as in [35–38], and other kinds of problems, such as diagnosis [39] and optimization [40].

The rest of the monograph is composed of seven chapters, which are shortly mentioned below. Chapter 2 deals with type-3 fuzzy concepts, which is an essential material to comprehend later chapters in the book. Chapter 3 presents an overview of type-3 fuzzy prediction works in the literature. Chapter 4 outlines the approach for COVID-19 prediction based on type-3 and fractal theory. Chapter 5 describes type-3 fuzzy aggregation of neural models and its use for financial forecasting. Chapter 6 delineates type-3 fuzzy aggregation for ensembles. Chapter 7 describes the optimization of type-3 systems and ensembles for prediction using the firefly algorithm. Finally, Chap. 8 outlines some conclusions about this work.

Lastly, we can state that this monograph will be a reference on type-3 for prediction that could serve as an inspiration to other researchers in contributing to this important new area of prediction.

References

1. P. Melin, O. Castillo, *Modelling, Simulation and Control of Non-Linear Dynamical Systems* (Taylor and Francis, London, Great Britain, 2002)
2. J.M. Mendel, Uncertainty, fuzzy logic, and signal processing. Signal Process. J. **80**, 913–933 (2000)
3. L.A. Zadeh, The concept of a linguistic variable and its application to approximate reasoning. Inf. Sci. **8**, 43–80 (1975)
4. J.R. Jang, C.T. Sun, E. Mizutani, *Neuro-Fuzzy and Soft Computing* (Prentice Hall, Upper Saddle River, NJ, USA, 1997)
5. O. Castillo, P. Melin, *Type-2 Fuzzy Logic: Theory and Applications* (Springer, Heidelberg, Germany, 2008)
6. N. N. Karnik, J. M. Mendel, An introduction to type-2 fuzzy logic systems. Technical Report, University of Southern California, 1998
7. M. Wagenknecht, K. Hartmann, Application of fuzzy sets of type 2 to the solution of fuzzy equations systems. Fuzzy Sets Syst. **25**, 183–190 (1988)
8. M.H.F. Zarandi, I.B. Turksen, O.T. Kasbi, Type-2 fuzzy modelling for desulphurization of steel process. Expert Syst. Appl. **32**, 157–171 (2007)

9. A. Mohammadzadeh, O. Castillo, S.S. Band et al., A novel fractional-order multiple-model type-3 fuzzy control for nonlinear systems with unmodeled dynamics. Int. J. Fuzzy Syst. **23**, 1633–1651 (2021)

10. H. Hagras, Hierarchical type-2 fuzzy logic control architecture for autonomous mobile robots. IEEE Trans. Fuzzy Syst. **12**, 524–539 (2004)

11. S. Coupland, R. John, New geometric inference techniques for type-2 fuzzy sets. Int. J. Approx. Reason. **49**, 198–211 (2008)

12. J.T. Starczewski, Efficient triangular type-2 fuzzy logic systems. Int. J. Approx. Reason. **50**, 799–811 (2009)

13. C. Walker, E. Walker, Sets with type-2 operations. Int. J. Approx. Reason. **50**, 63–71 (2009)

14. N.S. Bajestani, A. Zare, Application of optimized type-2 fuzzy time series to forecast Taiwan stock index, in *2nd International Conference on Computer, Control and Communication* (2009), pp. 275–280.

15. J.R. Castro, O. Castillo, P. Melin, A. Rodriguez-Diaz, A hybrid learning algorithm for a class of interval type-2 fuzzy neural networks. Inf. Sci. **179**, 2175–2193 (2009)

16. T. Dereli, A. Baykasoglu, K. Altun, A. Durmusoglu, I.B. Turksen, Industrial applications of type-2 fuzzy sets and systems: a concise review. Comput. Ind. **62**, 125–137 (2011)

17. C. Leal-Ramirez, O. Castillo, P. Melin, A. Rodriguez-Diaz, Simulation of the bird age-structured population growth based on an interval type-2 fuzzy cellular structure. Inf. Sci. **181**, 519–535 (2011)

18. R. Martinez, O. Castillo, L.T. Aguilar, Optimization of interval type-2 fuzzy logic controllers for a perturbed autonomous wheeled mobile robot using genetic algorithms. Inf. Sci. **179**(13), 2158–2174 (2009)

19. M. Hsiao, T.H.S. Li, J.Z. Lee, C.H. Chao, S.H. Tsai, Design of interval type-2 fuzzy sliding-mode controller. Inf. Sci. **178**(6), 1686–1716 (2008)

20. P. Melin, O. Castillo, A new method for adaptive model-based control of non-linear dynamic plants using a neuro-fuzzy-fractal approach. J. Soft. Comput. **5**, 171–177 (2001)

21. P. Melin, O. Castillo, A new method for adaptive model-based control of nonlinear plants using type-2 fuzzy logic and neural networks, in *Proceedings of IEEE FUZZ Conference* (2003), pp. 420–425

22. T. Ozen, J.M. Garibaldi, Investigating adaptation in type-2 fuzzy logic systems applied to umbilical acid-base assessment, in *European Symposium on Intelligent Technologies, Hybrid Systems and their Implementation on Smart Adaptive Systems (EUNITE 2003), Oulu, Finland* (2003)

23. R. Sepulveda, O. Castillo, P. Melin, O. Montiel, An efficient computational method to implement type-2 fuzzy logic in control applications. Adv. Soft Comput. **41**, 45–52 (2007)

24. R. Sepulveda, O. Castillo, P. Melin, A. Rodriguez-Diaz, O. Montiel, Experimental study of intelligent controllers under uncertainty using type-1 and type-2 fuzzy logic. Inf. Sci. **177**(10), 2023–2048 (2007)

25. P. Melin, O. Castillo, *Hybrid Intelligent Systems for Pattern Recognition* (Springer, Heidelberg, Germany, 2005)

26. O. Mendoza, P. Melin, O. Castillo, G. Licea, Type-2 fuzzy logic for improving training data and response integration in modular neural networks for image recognition. Lect. Notes Artif. Intell. **4529**, 604–612 (2007)

27. O. Mendoza, P. Melin, O. Castillo, Interval type-2 fuzzy logic and modular neural networks for face recognition applications. Appl. Soft Comput. J. **9**, 1377–1387 (2009)

28. O. Mendoza, P. Melin, G. Licea, Interval type-2 fuzzy logic for edges detection in digital images. Int. J. Intell. Syst. **24**, 1115–1133 (2009)

29. J. Urias, D. Hidalgo, P. Melin, O. Castillo, A method for response integration in modular neural networks with type-2 fuzzy logic for biometric systems. Adv. Soft Comput. **41**, 5–15 (2007)

30. P. Melin, O. Castillo, An intelligent hybrid approach for industrial quality control combining neural networks, fuzzy logic and fractal theory. Inf. Sci. **177**, 1543–1557 (2007)

31. O. Castillo, P. Melin, Hybrid intelligent systems for time series prediction using neural networks, fuzzy logic and fractal theory. IEEE Trans. Neural Netw. **13**, 1395–1408 (2002)

32. O. Castillo, P. Melin, A new fuzzy-fractal-genetic method for automated mathematical modelling and simulation of robotic dynamic systems, in *1998 IEEE International Conference on Fuzzy Systems (FUZZ-IEEE 1998) Proceedings*, vol. 2 (1998), pp. 1182–1187

33. O. Castillo, P. Melin, Intelligent adaptive model-based control of robotic dynamic systems with a hybrid fuzzy-neural approach. Appl. Soft Comput. **3**(4), 363–378 (2003)

34. P. Melin, O. Castillo, Adaptive intelligent control of aircraft systems with a hybrid approach combining neural networks, fuzzy logic and fractal theory. Appl. Soft Comput. **3**(4), 353–362 (2003)

35. O. Castillo, J.R. Castro, P. Melin, Interval type-3 fuzzy aggregation of neural networks for multiple time series prediction: the case of financial forecasting. Axioms **11**, 251 (2022). https://doi.org/10.3390/axioms11060251

36. M. Ramirez, P. Melin, A new perspective for multivariate time series decision making through a nested computational approach using type-2 fuzzy integration. Axioms **12**, 385 (2023). https://doi.org/10.3390/axioms12040385

37. M. Ramírez, P. Melin, O. Castillo, Interval type-3 fuzzy aggregation for hybrid-hierarchical neural classification and prediction models in decision-making. Axioms **12**, 906 (2023). https://doi.org/10.3390/axioms12100906

38. P. Melin, D. Sánchez, J.R. Castro, O. Castillo, Design of type-3 fuzzy systems and ensemble neural networks for COVID-19 time series prediction using a firefly algorithm. Axioms **11**, 410 (2022). https://doi.org/10.3390/axioms11080410

39. E. Ontiveros, P. Melin, O. Castillo, Comparative study of interval type-2 and general type-2 fuzzy systems in medical diagnosis. Inf. Sci. **525**, 37–53 (2020)

40. F. Valdez, J.C. Vazquez, P. Melin, O. Castillo, Comparative study of the use of fuzzy logic in improving particle swarm optimization variants for mathematical functions using co-evolution. Appl. Soft Comput. **52**, 1070–1083 (2017)

Chapter 2
Type-3 for Prediction

This chapter outlines type-3 concepts and their application in prediction.

2.1 Type-3 Fuzzy Sets

Fuzzy Logic has evolved from the original works of Zadeh with type-1 fuzzy theory [1–3], to later the developments of type-2 fuzzy theory and its applications [4–10], to now where type-3 fuzzy theory and its applications are emerging [11–16]. Even type-n has been envisioned [17]. We start by postulating the concepts.

Definition 2.1 A type-3 fuzzy set (T3 FS) [18], denoted by $A^{(3)}$, is represented by the plot of a function, called MF of $A^{(3)}$, in the Cartesian product $X \times [0, 1] \times [0, 1]$ in [0, 1], where X is the primary variable universe of $A^{(3)}$, x. The MF of $\mu_{A^{(3)}}$ is a type-3 MF (T3 MF):

$$\mu_{A^{(3)}} : X \times [0, 1] \times [0, 1] \rightarrow [0, 1]$$

$$A^{(3)} = \{(x, u(x), v(x, u), \mu_{A^{(3)}}(x, u, v)) | x \in X, u \in U \subseteq [0, 1], v \in V \subseteq [0, 1]\} \tag{2.1}$$

where U is universe for secondary variable u and V is universe for tertiary variable v. A is formulated as:

$$A^{(3)} = \int_{x \in X} \int_{u \in [0,1]} \int_{v \in [0,1]} \mu_{A^{(3)}}(x, u, v)/(x, u, v) \tag{2.2}$$

© The Author(s), under exclusive license to Springer Nature Switzerland AG 2024
O. Castillo and P. Melin, *Type-3 Fuzzy Logic in Time Series Prediction*,
SpringerBriefs in Computational Intelligence,
https://doi.org/10.1007/978-3-031-59714-5_2

$$A^{(3)} = \int_{x \in X} \left[\int_{u \in [0,1]} \left[\int_{v \in [0,1]} \mu_{A^{(3)}}(x, u, v)/v \right]/u \right]/x \qquad (2.3)$$

where $\iiint$ is the union over x, u, v values.

Equation (2.3) postulates a T3 FS MF as:

$$A^{(3)} = \int_{x \in X} \mu_{A_x^{(3)}}(u, v)/x$$

$$\mu_{A_x^{(3)}}(u, v) = \int_{u \in [0,1]} \mu_{A_{(x,u)}^{(3)}}(v)/u$$

$$\mu_{A_{(x,u)}^{(3)}}(v) = \int_{v \in [0,1]} \mu_{A^{(3)}}(x, u, v)/v$$

where $\mu_{A_x^{(3)}}(u, v)$ is the primary MF, $\mu_{A_x^{(3)}}(u, v)$ is the secondary MF and $\mu_{A_{(x,u)}^{(3)}}(v)$ is the tertiary MF.

If $\mu_{A^{(3)}}(x, u, v) = 1$, the T3 FS, $A^{(3)}$, is equivalent to an interval type-3 fuzzy set (IT3 FS) written as $\mathbb{A}$, , expressed by (2.4).

$$\mathbb{A} = \int_{x \in X} \left[\int_{u \in [0,1]} \left[\int_{v \in [\underline{\mu}_\mathbb{A}(x,u), \overline{\mu}_\mathbb{A}(x,u)]} 1/v \right]/u \right]/x \qquad (2.4)$$

where

$$\mu_{\mathbb{A}(x,u)}(v) = \int_{v \in [\underline{\mu}_\mathbb{A}(x,u), \overline{\mu}_\mathbb{A}(x,u)]} 1/v$$

$$\mu_{\mathbb{A}(x)}(u, v) = \int_{u \in [0,1]} \left[\int_{v \in [\underline{\mu}_\mathbb{A}(x,u), \overline{\mu}_\mathbb{A}(x,u)]} 1/v \right]/u$$

$$\mathbb{A} = \int_{x \in X} \mu_{\mathbb{A}(x)}(u, v)/x$$

Assuming that $v \in [\underline{\mu}_\mathbb{A}(x, u), \overline{\mu}_\mathbb{A}(x, u)]$ and the lower and upper MFs $\underline{\mu}_\mathbb{A}(x, u)$, $\overline{\mu}_\mathbb{A}(x, u)$ are general type-2 MFs (T2 MFs) on the plane (x, u), Eq. (2.4) is symplified to an interval type-3 MF (IT3 MF), $\tilde{\mu}_\mathbb{A}(x, u) \in [\underline{\mu}_\mathbb{A}(x, u), \overline{\mu}_\mathbb{A}(x, u)]$, defined by Eq. (2.5).

$$\mathbb{A} = \int\limits_{x \in X} \int\limits_{u \in [0,1]} \tilde{\mu}_{\mathbb{A}}(x, u)/(x, u) \tag{2.5}$$

where the lower T2 MF $\mu_{\underline{\mathbb{A}}}(x, u)$, is a subset of T2 MF $\overline{\mu}_{\mathbb{A}}(x, u)$, this is, $\mu_{\underline{\mathbb{A}}}(x, u) \subseteq \overline{\mu}_{\mathbb{A}}(x, u)$, then $\mu_{\underline{\mathbb{A}}}(x, u) \leq \overline{\mu}_{\mathbb{A}}(x, u)$, and as a consequence, an IT3 FS is postulated by two T2 FSs, one inferior $\underline{A}$ with T2 MF $\mu_{\underline{\mathbb{A}}}(x, u)$ and a superior $\overline{A}$, with T2 MF $\overline{\mu}_{\mathbb{A}}(x, u)$ as given by Eqs. (2.6) and (2.7) (Fig. 2.1).

$$\underline{A} = \int\limits_{x \in X} \int\limits_{u \in [0,1]} \mu_{\underline{\mathbb{A}}}(x, u)/(x, u) = \int\limits_{x \in X} \left[\int\limits_{u \in [0,1]} \underline{f}_x(u)/u \right]/x \tag{2.6}$$

$$\overline{A} = \int\limits_{x \in X} \int\limits_{u \in [0,1]} \overline{\mu}_{\mathbb{A}}(x, u)/(x, u) = \int\limits_{x \in X} \left[\int\limits_{u \in [0,1]} \overline{f}_x(u)/u \right]/x \tag{2.7}$$

where, the secondary MFs of $\underline{A}$ and $\overline{A}$ are T1 MFs postulated by the equations

$$\mu_{\underline{A}(x)}(u) = \int\limits_{u \in J_x} \underline{f}_x(u)/u \tag{2.8}$$

$$\mu_{\overline{A}(x)}(u) = \int\limits_{u \in J_x} \overline{f}_x(u)/u. \tag{2.9}$$

Scaled Gaussians Type-3 MFs are employed. This MF is written as, $\tilde{\mu}_{\mathbb{A}}(x, u) =$ ScaleGaussScaleGaussIT3MF, with Gaussian $FOU(\mathbb{A})$, with parameters $[\sigma, m]$ for the UMF and for LMF the parameters λ (LowerScale), ℓ (LowerLag) to form the $DOU = [\underline{\mu}(x), \overline{\mu}(x)]$. The vertical cuts $\mathbb{A}_{(x)}(u)$ form the $FOU(\mathbb{A})$, and are IT2 FSs with Gaussian IT2 MFs, $\mu_{\mathbb{A}(x)}(u)$ with parameters $[\sigma_u, m(x)]$ for UMF and for LMF the λ, and ℓ. The IT3 MF, $\tilde{\mu}_{\mathbb{A}}(x, u) =$ ScaleGaussScaleGaussIT3MF(x,{{[σ, m]}, λ, ℓ}) is formulated with expressions:

$$= exp\left[-\frac{1}{2}\left(\frac{x - m}{\sigma}\right)^2 \right] \tag{2.10}$$

$$\underline{u}(x) = \lambda \cdot exp\left[-\frac{1}{2}\left(\frac{x - m}{\sigma^*}\right)^2 \right] \tag{2.11}$$

where $\sigma^* = \sigma\sqrt{\frac{\ln(\ell)}{\ln(\varepsilon)}}$, ε is the machine epsilon. If $\ell = 0$, then $\sigma^* = \sigma$. Then $\overline{u}(x)$ and $\underline{u}(x)$ are the DOU limits. The range, $\delta(u)$ and radius, σ_u are:

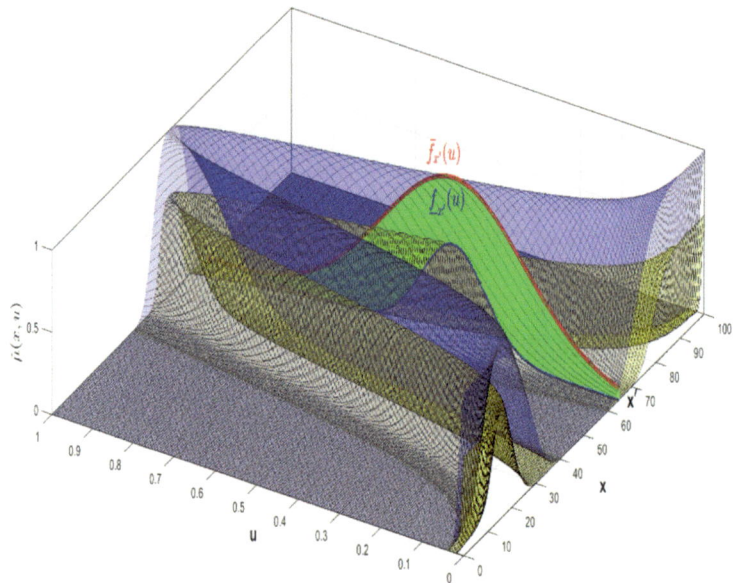

Fig. 2.1 IT3MF $\widetilde{\mu}\,(x,u)$ where $\underline{\mu}(x,u)$ is LMF and $\overline{\mu}(x,u)$ is UMF

$$\delta(u) = \overline{u}(x) - \underline{u}(x) \tag{2.12}$$

$$\sigma_u = \frac{\delta(u)}{2\sqrt{3}} + \varepsilon. \tag{2.13}$$

The apex or core, $m(x)$, of IT3 MF $\widetilde{\mu}\,(x,u)$, is given by:

$$m(x) = exp\left[-\frac{1}{2}\left(\frac{x-m}{\rho}\right)^2\right] \tag{2.14}$$

where $\rho = (\sigma + \sigma^*)/2$. The vertical cuts with IT2 MF, $\mu_{\mathbb{A}(x)}(u) = [\underline{\mu}_{\mathbb{A}(x)}(u), \overline{\mu}_{\mathbb{A}(x)}(u)]$, are expressed as:

$$\overline{\mu}_{\mathbb{A}(x)}(u) = exp\left[-\frac{1}{2}\left(\frac{u-u(x)}{\sigma_u}\right)^2\right] \tag{2.15}$$

$$\underline{\mu}_{\mathbb{A}(x)}(u) = \lambda \cdot exp\left[-\frac{1}{2}\left(\frac{u-u(x)}{\sigma_u^*}\right)^2\right] \tag{2.16}$$

where $\sigma_u^* = \sigma_u \sqrt{\frac{\ln(\ell)}{\ln(\varepsilon)}}$. If $\ell = 0$, then $\sigma_u^* = \sigma_u$. Then, $\overline{\mu}_{\mathbb{A}(x)}(u)$ and $\underline{\mu}_{\mathbb{A}(x)}(u)$ are the UMF and LMF of the secondary IT2MF vertical cuts.

2.2 Type-3 Fuzzy Systems

The IT3 system is postulated in the same way as the T2 FLSs.

Definition 2.2 The rule format is:

$$R_Z^k : IF\ x_1\ is\ \mathbb{F}_1^k\ and \ldots and\ x_i\ is\ \mathbb{F}_i^k\ and \ldots and\ x_n\ is\ \mathbb{F}_n^k$$
$$THEN\ y_1\ is\ \mathbb{G}_1^k,\ \ldots,\ y_j\ is\ \mathbb{G}_j^k,\ \ldots,\ y_m\ is\ \mathbb{G}_m^k$$

where i = 1,..., n, j = 1,...,m and k = 1,..., r, are numbers of inputs, outputs and rules, respectively.

We express the rule antecedents with a fuzzy relation $\mathbb{A}^k = \mathbb{F}_1^k \times \cdots \times \mathbb{F}_n^k$, utilizing the Cartesian product, $\times$, with IT3 FSs, $\mathbb{F}_i^k$, and the implication for the j-th output consequent, $\mathbb{G}_j^k$; then, rule $\mathbb{R}_j^k$ is formulated as:

$$\mathbb{R}_j^k = \mathbb{A}^k \rightarrow \mathbb{G}_j^k \qquad (2.17)$$

The n-dimensional input, is expressed by a type-2 relation, $\mathbb{A}_{\mathbb{X}'}$, as

$$\mathbb{A}_{\mathbb{X}'} = \mathbb{X}_1 \times \cdots \times \mathbb{X}_n \qquad (2.18)$$

Rule $\mathbb{R}_j^k$ postulates a fuzzy set of consequent $\mathbb{B}_j^k = \mathbb{A}_{\mathbb{X}'} \circ \mathbb{R}_j^k$ in **Y** such that

$$\mathbb{B}_j^k = \left[\mathbb{X}_1 \circ \left(\mathbb{F}_1^k \times \mathbb{G}_j^k\right)\right] \times \cdots \times \left[\mathbb{X}_n \circ \left(\mathbb{F}_n^k \times \mathbb{G}_j^k\right)\right] = \times_{i=1}^n \left[\mathbb{X}_i \circ \left(\mathbb{F}_i^k \times \mathbb{G}_j^k\right)\right] \qquad (2.19)$$

where the *rule activation level*, is an IT3 FS, $\mathbb{B}_j^k$. By aggregating all sets, $\mathbb{B}_j^k$, we find the aggregated set $\mathbb{B}_j$ for j = 1,...,m.

$$\mathbb{B}_j = \mathbb{B}_j^1 \cup \cdots \cup \mathbb{B}_j^k \cup \cdots \cup \mathbb{B}_j^r = \bigcup_{k=1}^r \mathbb{B}_j^k \qquad (2.20)$$

The expression $\hat{y}_j = f(x)$ is a IT3 model ($y_j is \mathbb{B}_j$), where sets $\mathbb{B}_j^k$ are submodels of $\mathbb{B}_j$.

Equation (2.21) is deduced from the relation, $\mu_{\mathbb{B}_j^k}(y_j|x')$, and is

$$\mu_{\mathbb{B}_j^k}(y_j|x') = \mu_{\mathbb{A}_{\mathbb{X}'} \circ \mathbb{R}_j^k}(y_j|x') = \underset{x \in X}{\underbrace{sup}}\left[\mu_{\mathbb{A}_{\mathbb{X}'}}(x) \sqcap \mu_{\mathbb{A}^k \rightarrow \mathbb{G}_j^k}(x, y_j)\right], y \in Y \qquad (2.21)$$

where $\mu_{\mathbb{B}_j^k}(y_j|\mathbf{x}')$ is the relation between the inference and output sets. The composition ($\circ$) is a mapping from input $\mathbf{x}'$ to an IT3 FS with MF, $\mu_{\mathbb{B}^k}(y_j|\mathbf{x}')$ ($y_j \in \mathbf{Y}$) of the output y_j. Simplifying Eq. (2.21), we arrive to

$$\mu_{\mathbb{B}_j^k}(y_j|\mathbf{x}') = \overset{\sim k}{\Phi}(\mathbf{x}') \sqcap \mu_{\mathbb{G}_j^k}(y_j) \tag{2.22}$$

where

$$\overset{\sim k}{\Phi}(\mathbf{x}') = \sqcap_{i=1}^n \left[\underbrace{sup}_{x_i \in X_i} \mu_{\mathbb{Q}_i^k}(x_i|x_i') \right] \tag{2.23}$$

$$\mu_{\mathbb{Q}_i^k}(x_i|x_i') = \mu_{\mathbb{X}_i}(x_i|x_i') \sqcap \mu_{\mathbb{F}_i^k}(x_i) \tag{2.24}$$

maximizing function $\mu_{\mathbb{Q}_i^k}(x_i|x_i')$, we obtain the supremum value in $x = x_{k,i}^{max}$,

$$x_{k,i}^{max} \equiv \underbrace{argmax}_{x_i} \left\{ \underbrace{sup}_{x_i \in X_i} \mu_{\mathbb{Q}_i^k}(x_i|x_i') \right\} \tag{2.25}$$

The *firing strength* $\overset{\sim k}{\Phi}(\mathbf{x}')$, is the membership of the t-norm operation, $\sqcap$ of all supreme membership values $\mu_{\mathbb{Q}_i^k}(x_{k,i}^{max}|x_i')$ of the intersection of each input $\mu_{\mathbb{X}_i}(x_i|x_i')$ with its antecedent $\mu_{\mathbb{F}_i^k}(x_i)$ that contribute to rule activation, i.e.,

$$\overset{\sim k}{\Phi}(\mathbf{x}') = \sqcap_{i=1}^n \mu_{\mathbb{Q}_i^k}(x_{k,i}^{max}|x_i') \tag{2.26}$$

The *rule activation level*, is the membership $\mu_{\mathbb{B}_j^k}(y_j|\mathbf{x}')$ resulting from the operation $\sqcap$ of the firing strength $\overset{\sim k}{\Phi}(\mathbf{x}')$ and the rule consequent $\mu_{\mathbb{G}_j^k}(y_j)$, this is, the composition operation ($\circ$) of the facts and rules described by the relation $\mathbb{B}_j^k = \mathbb{A}_{X'} \circ \mathbb{R}_j^k$.

Equation (2.27) for the MF of fuzzy relation, $\mu_{\tilde{B}_j}(y_j|\mathbf{x}')$, is the aggregation of all the rules for each output $j = 1,\ldots,m$, employing the operator *join* ($\sqcup$)—fuzzy union—.Rule combination employing the *join* is formulated by:

$$\mu_{\mathbb{B}_j}(y_j|\mathbf{x}') = \mu_{\mathbb{B}_j^1}(y_j|\mathbf{x}') \sqcup \cdots \sqcup \mu_{\mathbb{B}_j^k}(y_j|\mathbf{x}') \sqcup \cdots \sqcup \mu_{\mathbb{B}_j^r}(y_j|\mathbf{x}') = \sqcup_{k=1}^r \mu_{\mathbb{B}_j^k}(y_j|\mathbf{x}') \tag{2.27}$$

or

$$\mu_{\mathbb{B}_j}(y_j|\mathbf{x}') = \sqcup_{k=1}^r \left[\overset{\sim k}{\Phi}(\mathbf{x}') \sqcap \mu_{\mathbb{G}_j^k}(y_j) \right] \tag{2.28}$$

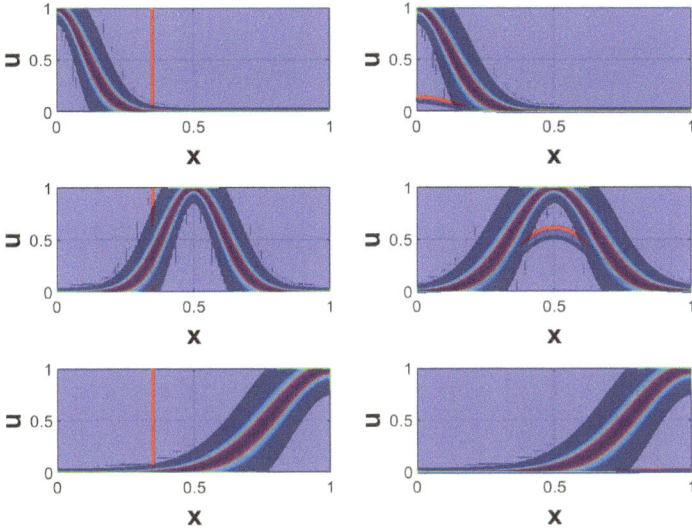

Fig. 2.2 Illustration of the inference process for a value of x = 0.35

For situations requiring a numeric output, $\mu_{\mathbb{B}_j}(y_j|\mathbf{x}')$ is converted to an IT2 FS, and then to a crisp value $\hat{y}_j$. Type-reduction methods are similar to type-2.

$$\hat{y}_j = typeReduction(y_j, \mu_{\mathbb{B}_j}(y_j|\mathbf{x}')) \tag{2.29}$$

In Fig. 2.2 the inference in a hypothetical system for x = 0.35 is illustrated, as well as its type-reduction in Fig. 2.3. In Fig. 2.4 a similar thing is done for x = 0.75 and in Fig. 2.5 the resulting reduction is found.

The structure of an interval type-3 system, is almost the same as for type-2 and type-1, and it is composed of a fuzzifier, rules, inference, type reduction and defuzzifier [20]. In Fig. 2.6 we can find the structure of an IT3 system [18, 19].

In some cases, the solution is to select the most important inputs, as in Fig. 2.7. We have used this hierarchical approach in several applications, like in [20] where three type-1 individual controllers of the airplane are combined with a type-2 fuzzy aggregator, but we have achieved other successful applications [21].

2.3 Summary

In this chapter we have reviewed the more important concepts of type-3 fuzzy logic theory, so that the reader can have the basic understanding of the fundamentals needed to outline in the following chapters the applications in prediction. On the theoretical side of this work we can mention that the presented theory could be extended to type-4 and type-n, we believe that this will an interesting avenue of theoretical research

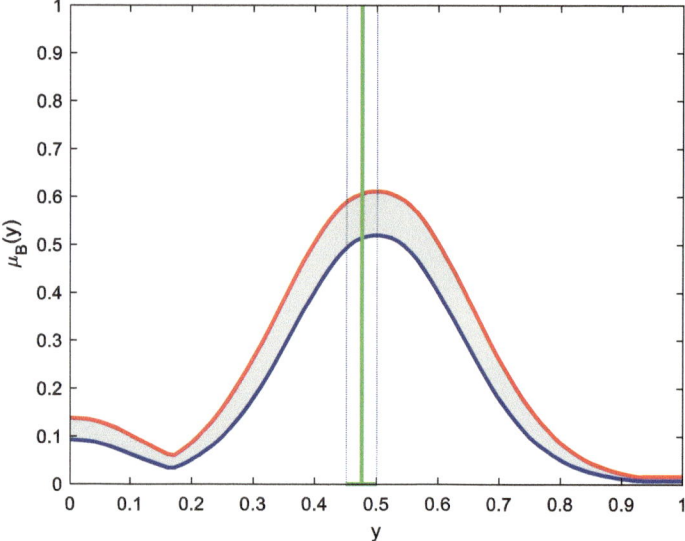

Fig. 2.3 Illustration of type reduction process for x = 0.35

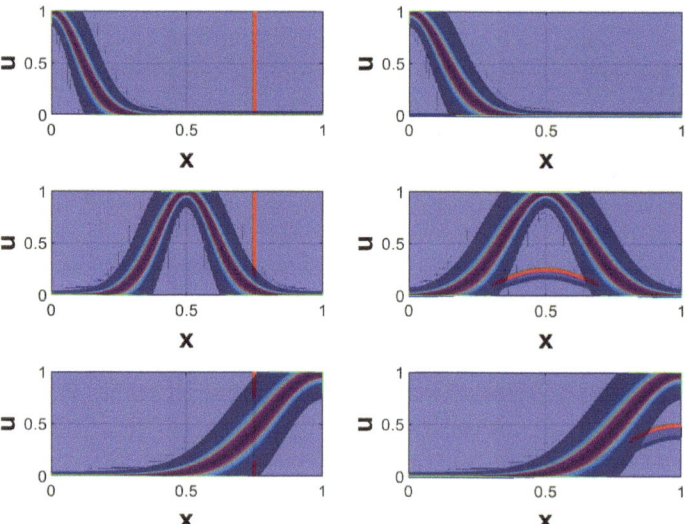

Fig. 2.4 Illustration of the inference process for x = 0.75

work in the future. Also, theoretically speaking hybrid type-3 concepts could be proposed, like type-3 hesitant sets, type-3 mediative sets, and others, that could be investigated to find out if other kinds of uncertainty could be handled in a better way [22].

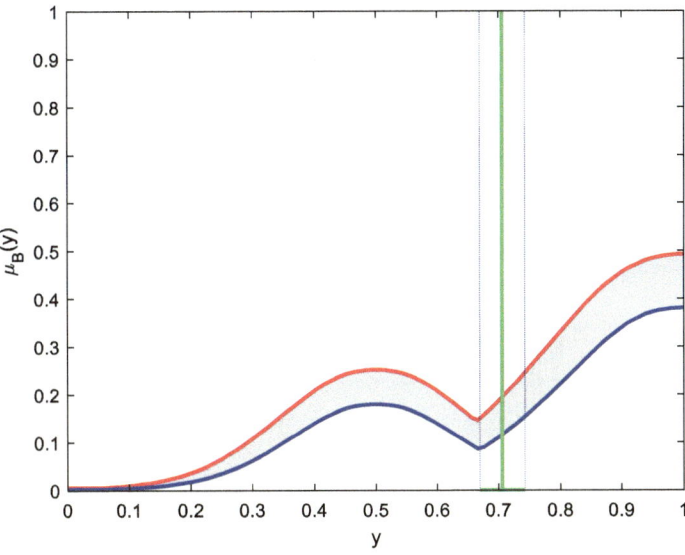

Fig. 2.5 Illustration of type reduction process for x = 0.75

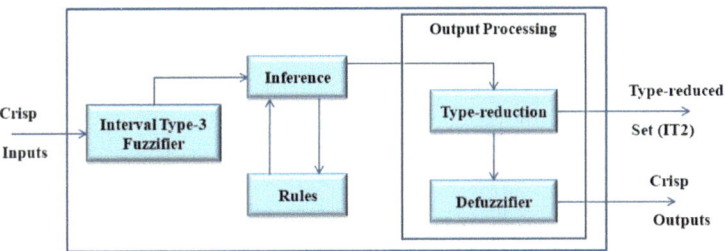

Fig. 2.6 Architecture of a type-3 system

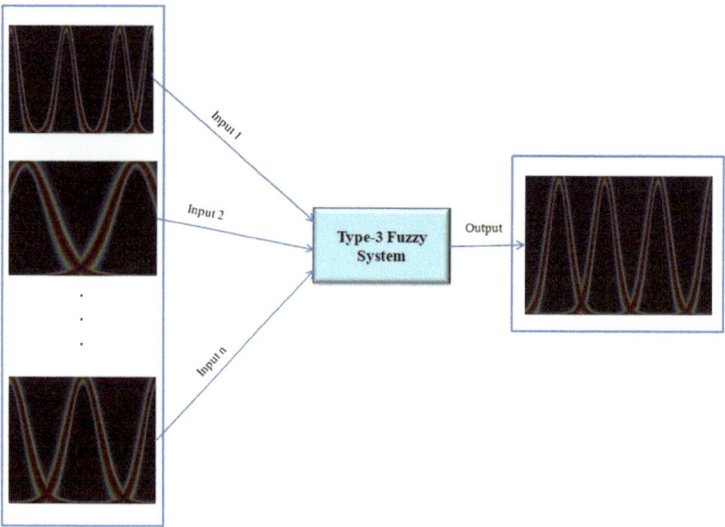

Fig. 2.7 Hierarchical approach

References

1. L.A. Zadeh, The concept of a linguistic variable and its application to approximate reasoning. Inf. Sci. **8**, 43–80 (1975)
2. L.A. Zadeh, Knowledge representation in fuzzy logic. IEEE Trans. Knowl. Data Eng. **1**, 89 (1989)
3. L.A. Zadeh, Fuzzy logic. Computer **1**(4), 83–93 (1998)
4. J.M. Mendel, H. Hagras, W.-W. Tan, W.W. Melek, H. Ying, *Introduction to Type-2 Fuzzy Logic Control* (Wiley and IEEE Press, Hoboken, 2014)
5. J.E. Moreno et al., Design of an interval Type-2 fuzzy model with justifiable uncertainty. Inf. Sci. **513**, 206–221 (2020)
6. F. Olivas, F. Valdez, O. Castillo, P. Melin, Dynamic parameter adaptation in particle swarm optimization using interval type-2 fuzzy logic. Soft. Comput. **20**(3), 1057–1070 (2016)
7. A. Sakalli, T. Kumbasar, J.M. Mendel, Towards systematic design of general type-2 fuzzy logic controllers: analysis, interpretation, and tuning. IEEE Trans. Fuzzy Syst. **29**(2), 226–239 (2021)
8. E. Ontiveros, P. Melin, O. Castillo, High order α-planes integration: a new approach to computational cost reduction of general type-2 fuzzy systems. Eng. Appl. Artif. Intell. **74**, 186–197 (2018)
9. E. Ontiveros, P. Melin, O. Castillo, Comparative study of interval type-2 and general type-2 fuzzy systems in medical diagnosis. Inf. Sci. **525**, 37–53 (2020)
10. J.R. Castro, O. Castillo, P. Melin, A. Rodriguez-Diaz, Building fuzzy inference systems with a new interval type-2 fuzzy logic toolbox. Trans. Comput. Sci. I 104–114
11. Y. Cao, A. Raise, A. Mohammadzadeh et al., Deep learned recurrent type-3 fuzzy system: application for renewable energy modeling / prediction. Energy Reports (2021)
12. A. Mohammadzadeh, O. Castillo, S.S. Band et al., A novel fractional-order multiple-model type-3 fuzzy control for nonlinear systems with unmodeled dynamics. Int. J. Fuzzy Syst. (2021). https://doi.org/10.1007/s40815-021-01058-1

13. S.N. Qasem, A. Ahmadian, A. Mohammadzadeh, S. Rathinasamy, B. Pahlevanzadeh, A type-3 logic fuzzy system: optimized by a correntropy based Kalman filter with adaptive fuzzy kernel size Inform. Sci. **572**, 424–443 (2021)
14. J.T. Rickard, J. Aisbett, G. Gibbon, Fuzzy subsethood for fuzzy sets of type-2 and generalized type-n. IEEE Trans. Fuzzy Syst. **17**(1), 50–60 (2009)
15. A. Mohammadzadeh, M.H. Sabzalian, W. Zhang, An interval type-3 fuzzy system and a new online fractional-order learning algorithm: Theory and practice. IEEE Trans. Fuzzy Syst. **28**(9), 1940–1950 (2020)
16. Z. Liu, A. Mohammadzadeh, H. Turabieh, M. Mafarja, S.S. Band, A. Mosavi, A new online learned interval type-3 fuzzy control system for solar energy management systems. IEEE Access **9**, 10498–10508 (2021)
17. O. Castillo, Towards finding the optimal n in designing type-n fuzzy systems for particular classes of problems: a review. Appl. Comput. Math. **17**(1), 3–9 (2018)
18. O. Castillo, J.R. Castro, P. Melin, *Interval type-3 fuzzy systems: theory and design* (Springer, Cham, Switzerland, 2022)
19. O. Castillo, P. Melin, Towards interval type-3 intuitionistic fuzzy sets and systems. Mathematics **10**, 4091 (2022). https://doi.org/10.3390/math10214091
20. L. Cervantes, O. Castillo, Type-2 fuzzy logic aggregation of multiple fuzzy controllers for airplane flight control. Inf. Sci. **324**, 247–256 (2015)
21. O. Castillo, L. Cervantes, J. Soria, M. Sanchez, J.R. Castro, A generalized type-2 fuzzy granular approach with applications to aerospace. Inf. Sci. **354**, 165–177 (2016)
22. O. Castillo, P. Melin, Proposal for mediative fuzzy control: from type-1 to type-3. Symmetry **2023**, 15 (1941). https://doi.org/10.3390/sym15101941

Chapter 3
Type-3 Fuzzy Logic in Time Series Prediction

This chapter offers a review of type-3 in prediction and forecasting.

3.1 Introduction

Fuzzy systems started with the work of Zadeh in 1965 [1], which was later called type-1 as was the original proposal in this area. Later, Zadeh also put forward type-2 fuzzy sets [2] as an enhanced methodology to handle real-world uncertainty. The area of type-2 had its real development after 2000 with the works of Mendel [3] and others, which have also achieved real applications [4]. More recently, type-3 has emerged as a promissory new area with real applicability, as in [5–15].

The idea is providing a review of type-3 for time series prediction. To fulfil this, a document search with "type-3 fuzzy" was made on Scopus. The documents represent a sample of the relevant type-3 works on the Scopus database [16–51]. After that, we also performed the search for documents with the keywords "type-3 fuzzy prediction or forecasting" to find out how many documents have been published in this area, and there are some interesting works in this area, like the ones presented in [52–58].

The remaining sections are: Sect. 3.2 contains a type-3 review. Section 3.3 offers a review of type-3 in prediction, and Sect. 3.4 offers a chapter summary.

3.2 Review of Type-3

We did undertake a search of papers containing the "Type-3 Fuzzy" words in the paper title. The publication total was of 65 up to June 30, 2023, as the study is constrained to publications with the words "type-3 fuzzy". First, in Fig. 3.1 we note how the citations have been increasing. Table 3.1 shows the citations per year (Y)

© The Author(s), under exclusive license to Springer Nature Switzerland AG 2024   17
O. Castillo and P. Melin, *Type-3 Fuzzy Logic in Time Series Prediction*,
SpringerBriefs in Computational Intelligence,
https://doi.org/10.1007/978-3-031-59714-5_3

from 2018. Figure 3.2 exhibits the documents on type-3 from 2010 and Table 3.2 lists papers per year from 2016. The papers with respect to the journal are exhibited in Fig. 3.3. A pictorial of documents per author is depicted in Fig. 3.4, while in Table 3.3 the papers per author are presented. Lastly, in Fig. 3.5 the papers per country (C) are presented and Table 3.4 summarizes the documents.

The paper distribution according to the journal is shown in Fig. 3.3.

Figure 3.3 highlights the Mathematics journal (MDPI publisher) as the one that has published more type-3 documents.

Fig. 3.1 Citations per year

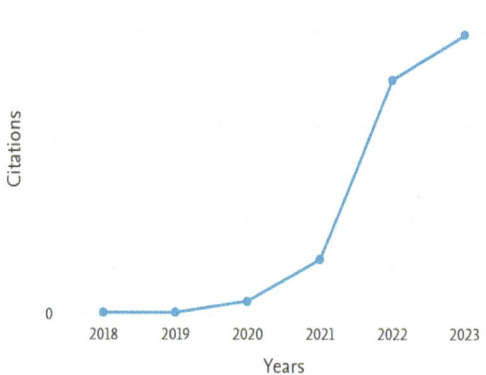

Table 3.1 Citations for type-3 papers

Y	2018	2019	2020	2021	2022	2023	Total
Citations	3	3	15	59	249	296	625

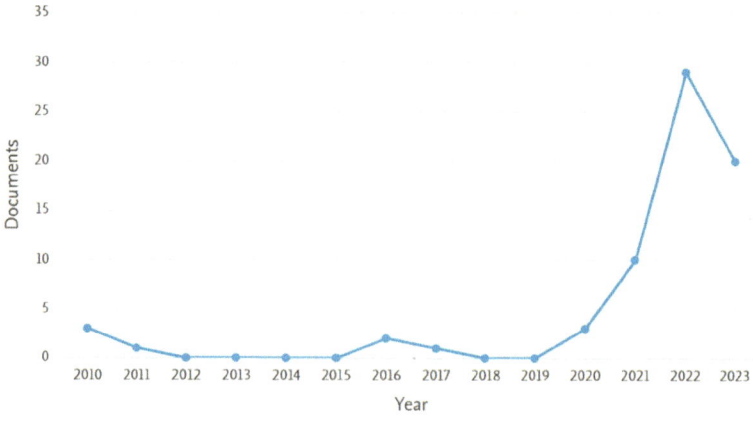

Fig. 3.2 Documents per year

Table 3.2 Type-3 papers from 2016

Y	2016	2017	2018	2019	2020	2021	2022	2023	Total
Papers	2	1	0	0	3	10	29	20	65

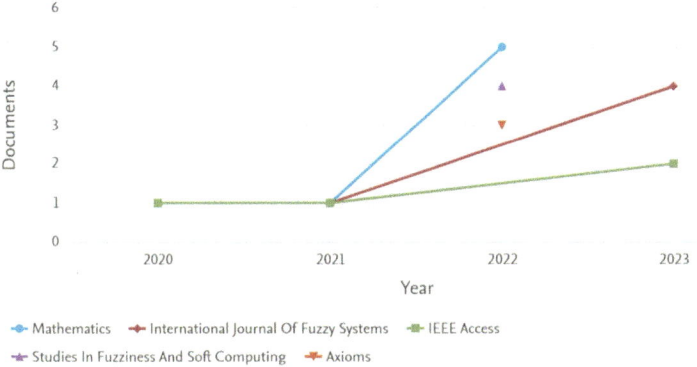

Fig. 3.3 Paper distribution according to the journal

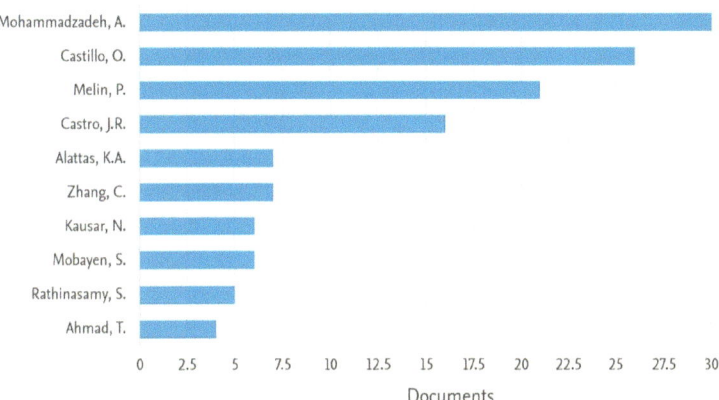

Fig. 3.4 Type-3 documents per author

Lastly, documents are classified by type (see Fig. 3.6), where it is noted that 81.2% of the total are documents in journals. Table 3.5 summarizes the number of documents per type. Figure 3.7 exhibits a categorization per area.

Table 3.3 Type-3 papers

Author	Papers
Mohammadzadeh, A	30
Castillo, O	26
Melin, P	21
Castro, JR	16
Alattas, KA	7
Zhang, C	7
Kausar, N	6
Mobayen, S	6
Rathinasamy, S	5
Ahmad, T	4

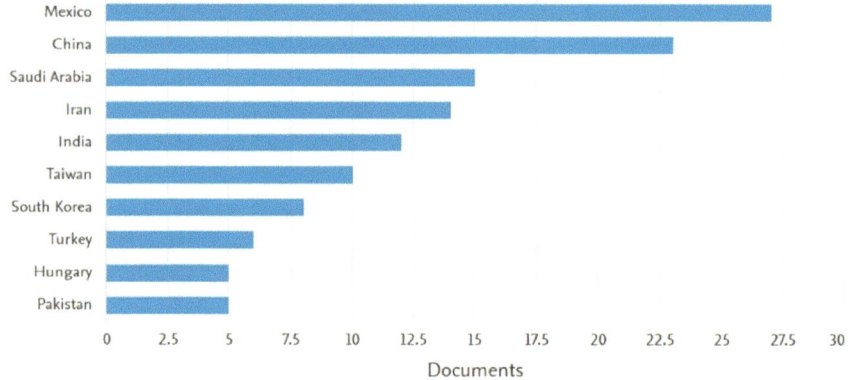

Fig. 3.5 Pictorial of documents per country

Table 3.4 Type-3 papers per country

C	Papers
Mexico	27
China	23
Saudi Arabia	15
Iran	14
India	12
Taiwan	10
South Korea	8
Turkey	6
Hungary	5
Pakistan	5

Fig. 3.6 Categorization
according to type

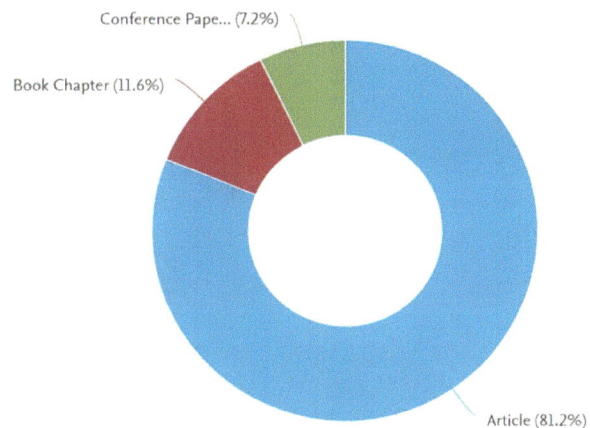

Table 3.5 Documents per
type

Type	Documents
Article	56
Chapter	8
Conference paper	5

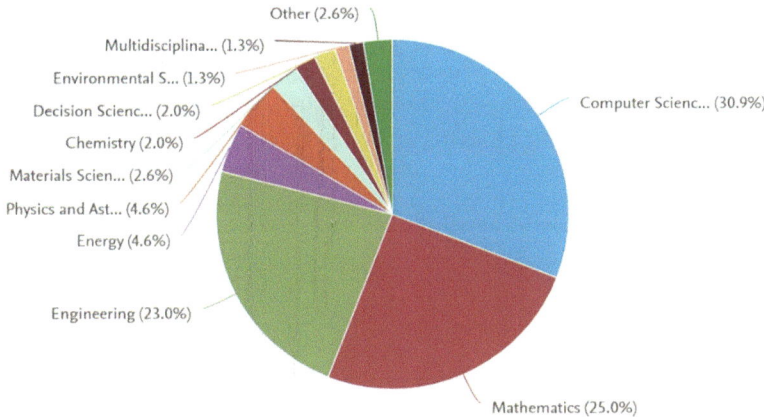

Fig. 3.7 Pictorial of documents per area

3.3   Type-3 in Prediction

We also did undertake a search with the words "Type-3 Fuzzy Prediction or Forecasting" in the document title, and we obtained 12 documents (December 30, 2023).
Figures and Tables were constructed with this search. Figure 3.8 shows the citations

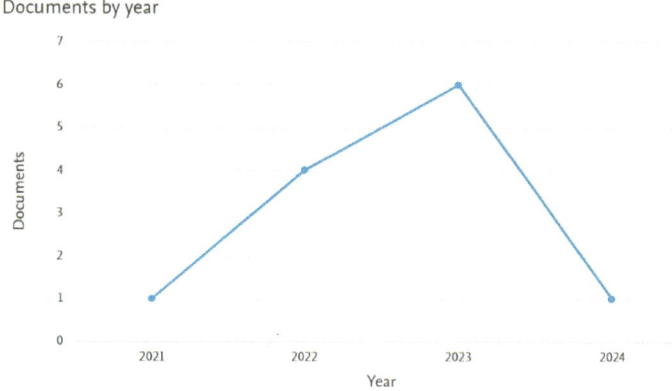

Fig. 3.8 Citations to type-3 prediction papers

to prediction papers in the period from 2019 to 2023. We offer in Table 3.6 the number of citations per year (Table 3.7).

We exhibit in Fig. 3.9 the number of papers per author of type-3 in prediction. Finally, we illustrate in Fig. 3.10 the number of publications per institution, where Tijuana Institute of Technology is seen in the top.

The documents per affiliation are shown in Fig. 3.10.

We also show in Fig. 3.11 the distribution of publications per country of type-3 applied in prediction.

We also illustrate in Fig. 3.12 the clusters of authors working in the applications of type-3 fuzzy logic in prediction. Finally, we illustrate in Fig. 3.13 the average publications per year for the clusters of authors.

Table 3.6 Citations per year to prediction

Y	2021	2022	2023	Total
Citations	2	34	84	120

Table 3.7 Publications per year in prediction

Y	2021	2022	2023	2024	Total
Papers	1	4	6	1	12

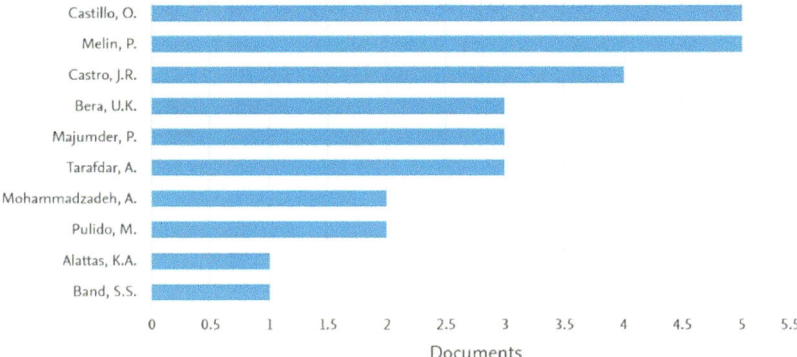

Fig. 3.9 Papers by author in prediction

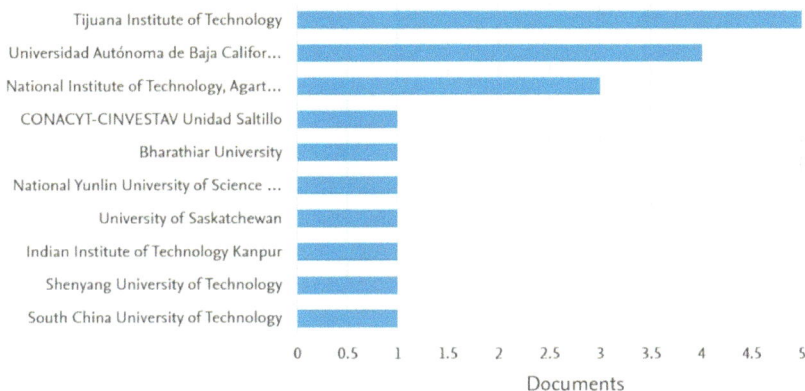

Fig. 3.10 Bar diagram with publications per affiliation

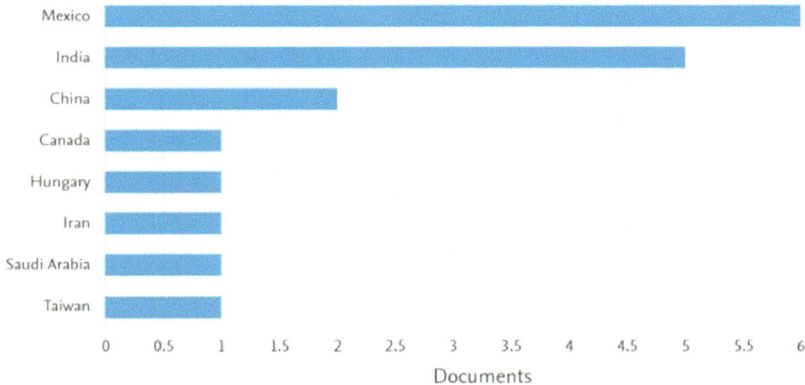

Fig. 3.11 Bar diagram with publications by affiliation for type-3 fuzzy logic in prediction

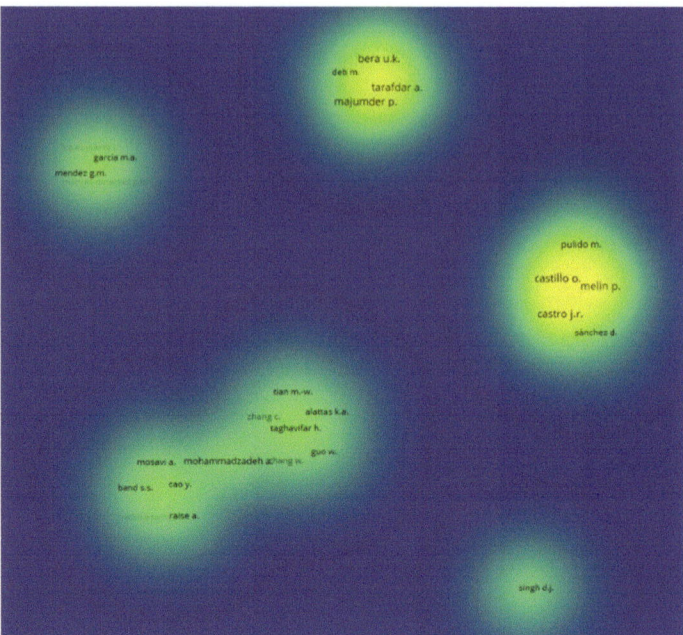

Fig. 3.12 Density plot showing clusters of authors

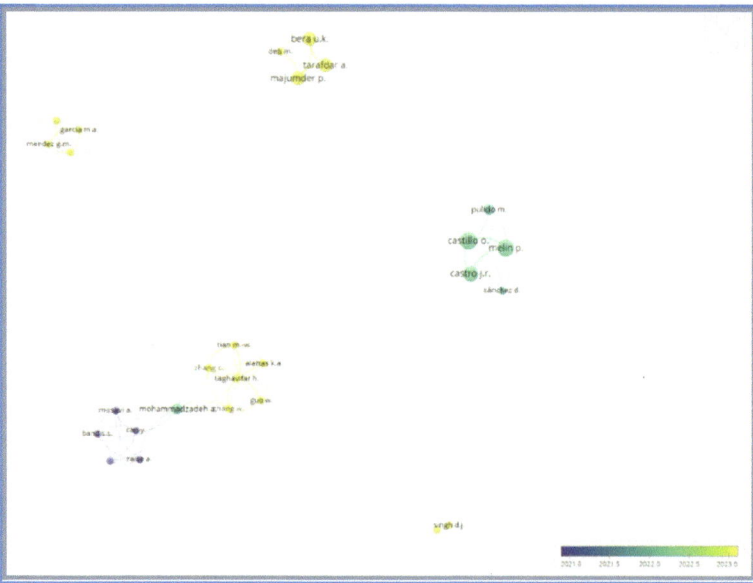

Fig. 3.13 Plot of clusters of authors according to the average publications per year

3.4 Summary

We have offered a review of the type-3 papers applied to prediction. We have mentioned the papers and citations that offer an overview of the achievements made in this area. We expect that future works for type-3 would be done along the lines of current works, such as in [59–67].

References

1. L.A. Zadeh, The concept of a linguistic variable and its application to approximate reasoning. Inf. Sci. **8**, 43–80 (1975)
2. L.A. Zadeh, Knowledge representation in fuzzy logic. IEEE Trans. Knowl. Data Eng. **1**, 89 (1989)
3. J.M. Mendel, H. Hagras, W.-W. Tan, W.W. Melek, H. Ying, *Introduction to Type-2 Fuzzy Logic Control* (Wiley and IEEE Press, Hoboken, 2014)
4. J.E. Moreno et al., Design of an interval type-2 fuzzy model with justifiable uncertainty. Inf. Sci. **513**, 206–221 (2020)
5. L. Amador-Angulo, O. Castillo, J.R. Castro, P. Melin, A new approach for interval type-3 fuzzy control of nonlinear plants. Int. J. Fuzzy Syst. **25**(4), 1624–1642 (2023)
6. P. Ochoa, O. Castillo, P. Melin, J.R. Castro, Interval type-3 fuzzy differential evolution for parameterization of fuzzy controllers. Int. J. Fuzzy Syst. **25**(4), 1360–1376 (2023)
7. O. Elhaki, K. Shojaei, A. Mohammadzadeh, S. Rathinasamy, Robust amplitude-limited interval type-3 neuro-fuzzy controller for robot manipulators with prescribed performance by output feedback. Neural Comput. Appl. **35**(12), 9115–9130 (2023)
8. O. Elhaki, K. Shojaei, A. Mohammadzadeh, Robust state and output feedback prescribed performance interval type-3 fuzzy reinforcement learning controller for an unmanned aerial vehicle with actuator saturation. IET Control Theory Appl. **17**(5), 605–627 (2023)
9. S. Xu, C. Zhang, A. Mohammadzadeh, Type-3 fuzzy control of robotic manipulators. Symmetry **15**(2), 2023
10. H. Huang, H. Xu, F. Chen, C. Zhang, A. Mohammadzadeh, An applied type-3 fuzzy logic system: practical matlab simulink and M-files for robotic, control, and modeling applications. Symmetry **15**(2) (2023)
11. O. Castillo, J.R. Castro, P. Melin, Forecasting the COVID-19 with interval type-3 fuzzy logic and the fractal dimension. Int. J. Fuzzy Syst. **25**(1), 182–197 (2023)
12. P. Melin, O. Castillo, Interval type-3 fuzzy decision making in quality evaluation for speaker manufacturing. Stud. Comput. Intell. **1096**, 489–498 (2023)
13. O. Castillo, P. Melin, Interval type-3 fuzzy decision making in material surface quality control. Stud. Comput. Intell. **1096**, 479–487 (2023)
14. A.S. Alkabaa, O. Taylan, M. Balubaid, C. Zhang, A. Mohammadzadeh, A practical type-3 fuzzy control for mobile robots: predictive and Boltzmann-based learning. Complex Intell. Syst. (2023).
15. B. Yan, X. Jiang, K.A. Alattas, C. Zhang, A. Mohammadzadeh, Generation of limit cycles in nonlinear systems: machine leaning based type-3 fuzzy control. IEEE Access **11**, 34835–34845 (2023)
16. A. Tarafdar, P. Majumder, M. Deb, U.K. Bera, Performance-emission optimization in a single cylinder CI-engine with diesel hydrogen dual fuel: a spherical fuzzy MARCOS MCGDM based type-3 fuzzy logic approach. Int. J. Hydrogen Energy (2023)
17. C. Peraza, O. Castillo, P. Melin, J.R. Castro, J.H. Yoon, Z.W. Geem, A type-3 fuzzy parameter adjustment in harmony search for the parameterization of fuzzy controllers. Int. J. Fuzzy Syst. (2023)

18. O. Castillo, J.R. Castro, P. Melin, Interval type-3 fuzzy systems: a natural evolution from type-1 and type-2 fuzzy systems. Stud. Comput. Intell. **1061**, 209–221 (2023)
19. M. Hamdy, A. Ibrahim, B. Abozalam, S. Helmy, Maximum power point tracking for solar photovoltaic system based on interval type-3 fuzzy logic: practical validation. Electr. Power Comput. Syst. **51**(10), 1009–1026 (2023)
20. A. Taghieh, A. Mohammadzadeh, C. Zhang, S. Rathinasamy, S. Bekiros, A novel adaptive interval type-3 neuro-fuzzy robust controller for nonlinear complex dynamical systems with inherent uncertainties. Nonlinear Dyn. **111**(1), 411–425 (2023)
21. A. Taghieh, C. Zhang, K.A. Alattas, Y. Bouteraa, S. Rathinasamy, A. Mohammadzadeh, A predictive type-3 fuzzy control for underactuated surface vehicles. Ocean Eng. **266** (2022)
22. M. Gheisarnejad, A. Mohammadzadeh, M. Khooban, Model predictive control based type-3 fuzzy estimator for voltage stabilization of DC power converters. IEEE Trans. Ind. Electron. **69**(12), 13849–13858 (2022)
23. O. Castillo, P. Melin, Towards interval type-3 intuitionistic fuzzy sets and systems. Mathematics **10**(21) (2022)
24. A. Taghieh, A. Mohammadzadeh, C. Zhang, N. Kausar, O. Castillo, A type-3 fuzzy control for current sharing and voltage balancing in microgrids. Appl. Soft Comput. (2022)
25. O. Castillo, J.R. Castro, P. Melin, Interval type-3 fuzzy fractal approach in sound speaker quality control evaluation. Eng. Appl. Artif. Intell. 116 (2022)
26. C. Peraza, P. Ochoa, O. Castillo, Z.W. Geem, Interval-type 3 fuzzy differential evolution for designing an interval-type 3 fuzzy controller of a unicycle mobile robot. Mathematics **10**(19) (2022)
27. O. Castillo, J.R. Castro, P. Melin, A methodology for building interval type-3 fuzzy systems based on the principle of justifiable granularity. Int. J. Intell. Syst. **37**(10), 7909–7943 (2022)
28. L. Amador-Angulo, O. Castillo, P. Melin, J.R. Castro, Interval type-3 fuzzy adaptation of the bee colony optimization algorithm for optimal fuzzy control of an autonomous mobile robot. Micromachines **13**(9) (2022)
29. G. Hua, F. Wang, J. Zhang, K.A. Alattas, A. Mohammadzadeh, V.M. The, A new type-3 fuzzy predictive approach for mobile robots. Mathematics **10**(17) (2022)
30. O. Castillo, J.R. Castro, M. Pulido, P. Melin, Interval type-3 fuzzy aggregators for ensembles of neural networks in COVID-19 time series prediction. Eng. Appl. Artif. Intell. 114 (2022)
31. D. Singh, N. Verma, A. Ghosh, A. Malagaudanavar, An approach towards the design of interval type-3 T-S fuzzy system. IEEE Trans. Fuzzy Syst. **30**(9), 3880–3893 (2022)
32. P. Melin, D. Sánchez, J.R. Castro, O. Castillo, Design of type-3 fuzzy systems and ensemble neural networks for COVID-19 time series prediction using a firefly algorithm. Axioms **11**(8) (2022)
33. M. Tian, S. Yan, J. Liu, K.A. Alattas, A. Mohammadzadeh, M.T. Vu, A new type-3 fuzzy logic approach for chaotic systems: robust learning algorithm. Mathematics **10**(15) (2022)
34. V. Kreinovich, O. Kosheleva, P. Melin, O. Castillo, Efficient algorithms for data processing under type-3 (and higher) fuzzy uncertainty. Mathematics **10**(13) (2022)
35. O. Castillo, J.R. Castro, P. Melin, Interval type-3 fuzzy control for automated tuning of image quality in televisions. Axioms **11**(6) (2022)
36. O. Castillo, J.R. Castro, P. Melin, Interval type-3 fuzzy aggregation of neural networks for multiple time series prediction: the case of financial forecasting. Axioms **11**(6), 251 (2022)
37. M. Gheisarnejad, A. Mohammadzadeh, H. Farsizadeh, M. Khooban, Stabilization of 5G telecom converter-based deep type-3 fuzzy machine learning control for telecom applications. IEEE Trans. Circuits Syst. Express Briefs **69**(2), 544–548 (2022)
38. A. Riaz, S. Kousar, N. Kausar, D. Pamucar, G.M. Addis, Codes over lattice-valued intuitionistic fuzzy set type-3 with application to the complex DNA analysis. Complexity (2022)
39. W. Fan, A. Mohammadzadeh, N. Kausar, D. Pamucar, N.A.D. Ide, A new type-3 fuzzy PID for energy management in microgrids. Adv. Math. Phys. (2022)
40. O. Castillo, M. Pulido, P. Melin, Interval type-3 fuzzy aggregators for ensembles of neural networks in time series prediction. Lect Notes Netw. Syst. (LNNS) **504**, 785–793 (2022)

41. M. Tian, Y. Bouteraa, K.A. Alattas, S. Yan, A.K. Alanazi, A. Mohammadzadeh, S. Mobayen, A type-3 fuzzy approach for stabilization and synchronization of chaotic systems: applicable for financial and physical chaotic systems. Complexity (2022)

42. O. Castillo, J.R. Castro, P. Melin, Introduction to interval type-3 fuzzy systems. Stud. Fuzziness Soft Comput. **418**, 1–4 (2022)

43. O. Castillo, J.R. Castro, P. Melin, Interval type-3 fuzzy logic systems (IT3FLS). Stud. Fuzziness Soft Comput. **418**, 45–98 (2022)

44. O. Castillo, J.R. Castro, P. Melin, Interval type-3 fuzzy sets. Stud. Fuzziness Soft Comput. **418**, 13–43 (2022)

45. A. Taghieh, A.A. Aly, B.F. Felemban, A. Althobaiti, A. Mohammadzadeh, A. Bartoszewicz, A hybrid predictive type-3 fuzzy control for time-delay multi-agent systems. Electronics (Switzerland) **11**(1) (2022)

46. M. Tian, A. Mohammadzadeh, J. Tavoosi, S. Mobayen, J.H. Asad, O. Castillo, A.R. Várkonyi-Kóczy, A deep-learned type-3 fuzzy system and its application in modeling problems. Acta Polytech Hung **19**(2), 151–172 (2022)

47. S. Yan, A.A. Aly, B.F. Felemban, M. Gheisarnejad, M. Tian, M.H. Khooban, A. Mohammadzadeh, S. Mobayen, A new event-triggered type-3 fuzzy control system for multi-agent systems: optimal economic efficient approach for actuator activating. Electronics (Switzerland) **10**(24) (2021)

48. A.A. Aly, B.F. Felemban, A. Mohammadzadeh, O. Castillo, A. Bartoszewicz, Frequency regulation system: a deep learning identification, type-3 fuzzy control and lmi stability analysis. Energies **14**(22) (2021)

49. M. Tian, S. Yan, A. Mohammadzadeh, J. Tavoosi, S. Mobayen, R. Safdar, W. Assawinchaichote, M.T. Vu, A. Zhilenkov, Stability of interval type-3 fuzzy controllers for autonomous vehicles. Mathematics **9**(21) (2021)

50. R.H. Vafaie, A. Mohammadzadeh, M.J. Piran, A new type-3 fuzzy predictive controller for MEMS gyroscopes. Nonlinear Dyn. **106**(1), 381–403 (2021)

51. A. Mohammadzadeh, O. Castillo, S.S. Band, A. Mosavi, A novel fractional-order multiple-model type-3 fuzzy control for nonlinear systems with unmodeled dynamics. Int. J. Fuzzy Syst. **23**(6), 1633–1651 (2021)

52. G.M. Mendez, I. Lopez-Juarez, P.N. Montes-Dorantes, M.A. Garcia, A new method for the design of interval type-3 fuzzy logic systems with uncertain type-2 non-singleton inputs (IT3 NSFLS-2): a case study in a hot strip mill. IEEE Access **11**, 44065–44081 (2023)

53. D.J. Singh, N.K. Verma, Interval type-3 T-S fuzzy system for nonlinear aerodynamic modeling. Appl. Soft Comput. **150**, 111097 (2024)

54. A. Tarafdar, P. Majumder, U.K. Bera, Prediction of air quality index in Kolkata city using an advanced learned interval type-3 fuzzy logic system, in *2023 IEEE 8th International Conference for Convergence in Technology, I2CT 2023, Lonavla, India, 2023* (2023), pp. 1–7

55. A. Tarafdar, P. Majumder, U.K. Bera, An advanced learned type-3 fuzzy logic-based hybrid system to optimize inventory cost for a new business policy. Proc. Natl. Acad. Sci. India Sect.-Phys. Sci. **93**(4), 711–727 (2023)

56. A. Tarafdar, P. Majumder, M. Deb, U.K. Bera, Application of a q-rung orthopair hesitant fuzzy aggregated Type-3 fuzzy logic in the characterization of performance-emission profile of a single cylinder CI-engine operating with hydrogen in dual fuel mode. Energy **269**, 126751 (2023)

57. M.-W. Tian, K.A. Alattas, W. Guo, H. Taghavifar, A. Mohammadzadeh, W. Zhang, C. Zhang, *A Strong Secure Path Planning/Following System Based on Type-3 Fuzzy Control, Multi-Switching Chaotic Systems, and Random Switching Topology* (Article in Press, Complex and Intelligent Systems, 2023)

58. Y. Cao, A. Raise, A. Mohammadzadeh, S. Rathinasamy, S.S. Band, A. Mosavi, Deep learned recurrent type-3 fuzzy system: application for renewable energy modeling/prediction. Energy Rep. **7**, 8115–8127 (2021)

59. F. Valdez, J.C. Vazquez, P. Melin, O. Castillo, Comparative study of the use of fuzzy logic in improving particle swarm optimization variants for mathematical functions using co-evolution. Appl. Soft Comput. **52**, 1070–1083 (2017)

60. D. Sánchez, P. Melin, O. Castillo, Comparison of particle swarm optimization variants with fuzzy dynamic parameter adaptation for modular granular neural networks for human recognition. J. Intell. Fuzzy Syst. **38**(3), 3229–3252 (2020)
61. R.-E. Precup, R.-C. David, R.-C. Roman, E.M. Petriu, A.-I. Szedlak-Stinean, Slime mould algorithm-based tuning of costeffective fuzzy controllers for servo systems. Int. J. Comput. Intell. Syst. **14**(1), 1042–1052 (2021)
62. R.-E. Precup, R.-C. David, R.-C. Roman, A.-I. Szedlak-Stinean, E.M. Petriu, Optimal tuning of interval type-2 fuzzy controllers for nonlinear servo systems using slime mould algorithm. Int. J. Syst. Sci. (2021). https://doi.org/10.1080/00207721.2021.1927236
63. D. Sanchez, P. Melin, O. Castillo, A grey wolf optimizer for modular granular neural networks for human recognition. Comput. Intell. Neurosci. 4180510:1–4180510:26 (2017)
64. O. Castillo, P. Melin, A new fuzzy-fractal-genetic method for automated mathematical modelling and simulation of robotic dynamic systems, in *1998 IEEE International Conference on Fuzzy Systems (FUZZ-IEEE 1998) Proceedings*, vol. 2, pp. 1182–1187
65. O. Castillo, P. Melin, Intelligent adaptive model-based control of robotic dynamic systems with a hybrid fuzzy-neural approach. Appl. Soft Comput. **3**(4), 363–378 (2003)
66. P. Melin, O. Castillo, Adaptive intelligent control of aircraft systems with a hybrid approach combining neural networks, fuzzy logic and fractal theory. Appl. Soft Comput. **3**(4), 353–362 (2003)
67. L. Aguilar, P. Melin, O. Castillo, Intelligent control of a stepping motor drive using a hybrid neuro-fuzzy ANFIS approach. Appl. Soft Comput. **3**(3), 209–219 (2003)

Chapter 4
Prediction with a Hybrid Interval Type-3 Fuzzy-Fractal Approach

Abstract In this chapter an approach for prediction with a hybrid of type-3 and fractal theories is offered.

4.1 Introduction

In this chapter the method is described. The fractal dimension [1] is utilized to measure the complexity of the time series. Based on values for the fractal dimensions of different time series, linguistic values for the dimensions can be constructed and then the fuzzy rules that can predict confirmed cases and deaths for the countries based on the complexity of a time series [2]. The fuzzy rules can be elucidated by employing fuzzy clustering on the data [3]. The key goal of employing type-3 is because it should be capable of surpassing type-2 in dealing with uncertainty in real problems.

The rules are outlined with the Mamdani reasoning method, and the centroid as the defuzzification approach [4]. However, Sugeno reasoning in which the consequents are mathematical functions is also possible [5]. The Sugeno fuzzy model can be constructed with a neuro-fuzzy approach [6] to learn from real data the ideal parameter values for the functions and for the MFs [7].

Recently, the rapid COVID-19 propagation has been noticed and spreading to all the world. In particular, in the case of Europe several countries, like Italy, Spain, and France have been hit very hard with the spread of the COVID-19 virus [8–13]. In the case of the American continent, United States, Canada and Brazil have also suffered a significant number of cases [14–17]. In summary, it is very important that research work should be undertaken for understanding all facets of this problem [18–20]. There are also some recent works on studying COVID-19 dynamics in space and time that inspired this work [21–28].

In the literature of intelligent models for prediction we can find neural networks (NNs) [6], ensemble NNs [28], adaptive neuro-fuzzy inference systems (ANFISs) [6], type-1 fuzzy systems (T1Fs) [4], type-2 fuzzy systems (T2FSs), and interval type-3 fuzzy systems (IT3FSs). We propose a mixing of IT3FS with the fractal dimension

O. Castillo and P. Melin, *Type-3 Fuzzy Logic in Time Series Prediction*,
SpringerBriefs in Computational Intelligence,
https://doi.org/10.1007/978-3-031-59714-5_4

(FD) that is called IT3FS + FD as it mixes type-3 and FD. The contribution of the chapter is the utilization of type-3 and the FD for prediction.

The other sections of the chapter are structured as: Sect. 4.2 outlines a background of fractal concepts. Section 4.3 is dedicated to type-3 terminology. Section 4.4 offers the proposed approach. Section 4.5 summarizes the results. Lastly, Sect. 4.6 is dedicated to the conclusions of the chapter.

4.2   Fractal Dimension

Recently, significant progress has been made in comprehending the complexity of an object through the utilization of fractal constructs [1]. For example, time series in finance and economics exhibit properties suggesting a fractal structure [29, 30]. The FD is formulated as:

$$d = \lim_{r \to 0} [lnN(r)]/[ln(1/r)] \qquad (4.1)$$

where $N(r)$ is for the number of boxes of size r. The d in (4.1) is approximated with box covering for several r sizes and then using regression for estimating the d value. The FD is estimated by:

$$lnN(r) = ln\beta - dlnr \qquad (4.2)$$

where d is the dimension. This regression is performed by employing as data a set of ordered pairs of number of boxes corresponding to different box sizes.

4.3   Type-3 Concepts

A fuzzy system can be employed as a prediction tool. In this situation we can begin by utilizing fuzzy clustering [3, 31, 32] to group the data, and then after that build a fuzzy system that will constitute a forecasting scheme for an application. Originally fuzzy sets and logic were put forward by Lotfi Zadeh in 1965 [33], which are now called type-1 fuzzy sets and logic, later himself proposed the type-2 fuzzy term in 1975 to better represent the uncertainty in the real world [4], and correspondingly has had recently many successful applications.

Definition 4.1 A type-3 fuzzy set (T3 FS) [34–36], denoted by $A^{(3)}$, is represented by the plot of a trivariate function, called membership function (MF) of $A^{(3)}$, in the Cartesian product $X \times [0, 1] \times [0, 1]$ in [0, 1], where X is the universe of the primary variable of $A^{(3)}$, x. The MF of $\mu_{A^{(3)}}$ is denoted by $\mu_{A^{(3)}}(x, u, v)$ (or $\mu_{A^{(3)}}$) and it is called a T3 MF of the T3 FS:

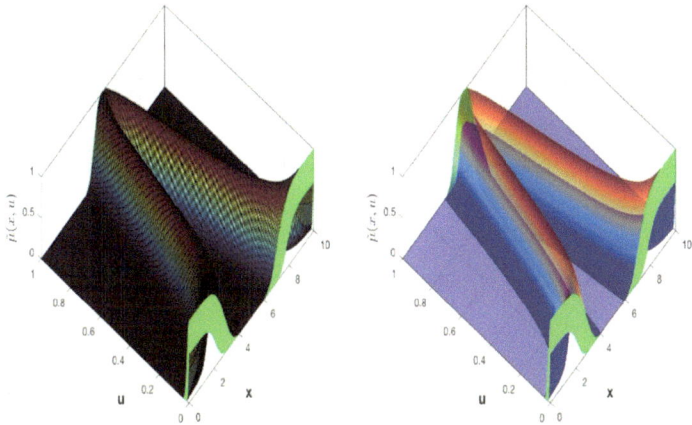

Fig. 4.1 Plot of the MFs of the IT3 FS

$$\mu_{A^{(3)}} : X \times [0, 1] \times [0, 1] \rightarrow [0, 1]$$

$$A^{(3)} = \{(x, u(x), v(x, u), \mu_{A^{(3)}}(x, u, v)) | x \in X, u \in U \subseteq [0, 1], v \in V \subseteq [0, 1]\} \tag{4.3}$$

where U is the universe of the secondary variable u and V is the universe for tertiary variable v. The 3D plot of the IT3MF is an isosurface formed by secondary IT2MFs $\mu_{\mathbb{A}(x)}(u)$ in green color in Fig. 4.1, which constitute the domain of uncertainty (DOU).

4.3.1 Type-3 Gaussian MFs

Type-3 MFs that are scaled Gaussians, which is, $\tilde{\mu}_{\mathbb{A}}(x, u) = $ ScaleGaussScaleGauss IT3MF, with Gaussian $FOU(\mathbb{A})$, with upper parameters $[\sigma, m]$ for the UMF and λ, ℓ or the LMF to form the $DOU = [\mu(x), \overline{\mu}(x)]$. The vertical cuts $\mathbb{A}_{(x)}(u)$ form the $FOU(\mathbb{A})$, and are Gaussian $\overline{IT}2$ MFs, $\mu_{\mathbb{A}(x)}(u)$ with parameters $[\sigma_u, m(x)]$ for UMF and LMF with λ, ℓ. The IT3 MF, $\tilde{\mu}_{\mathbb{A}}(x, u) = $ ScaleGaussScaleGaussIT3MF(x,{{[σ, m]}, λ, ℓ}) is formulated with expressions:

$$\overline{u}(x) = exp\left[-\frac{1}{2}\left(\frac{x - m}{\sigma}\right)^2\right] \tag{4.4}$$

$$\underline{u}(x) = \lambda \cdot exp\left[-\frac{1}{2}\left(\frac{x - m}{\sigma^*}\right)^2\right] \tag{4.5}$$

where $\sigma^* = \sigma\sqrt{\frac{\ln(\ell)}{\ln(\varepsilon)}}$, ε is the machine epsilon. If $\ell = 0$, then $\sigma^* = \sigma$. Then $\bar{u}(x)$ and $\underline{u}(x)$ are the DOU upper and lower limits. The range, $\delta(u)$ and radius, σ_u are:

$$\delta(u) = \bar{u}(x) - \underline{u}(x) \tag{4.6}$$

$$\sigma_u = \frac{\delta(u)}{2\sqrt{3}} + \varepsilon \tag{4.7}$$

The apex or core, $m(x)$, of the IT3 MF $\tilde{\mu}\ (x, u)$, is postulated by:

$$m(x) = exp\left[-\frac{1}{2}\left(\frac{x - m}{\rho}\right)^2\right] \tag{4.8}$$

where $\rho = (\sigma + \sigma^*)/2$. Then, the vertical cuts with IT2 MF, $\mu_{\mathbb{A}(x)}(u) = [\underline{\mu}_{\mathbb{A}(x)}(u), \overline{\mu}_{\mathbb{A}(x)}(u)]$, are described by equations:

$$\overline{\mu}_{\mathbb{A}(x)}(u) = exp\left[-\frac{1}{2}\left(\frac{u - u(x)}{\sigma_u}\right)^2\right] \tag{4.9}$$

$$\underline{\mu}_{\mathbb{A}(x)}(u) = \lambda \cdot exp\left[-\frac{1}{2}\left(\frac{u - u(x)}{\sigma_u^*}\right)^2\right] \tag{4.10}$$

where $\sigma_u^* = \sigma_u\sqrt{\frac{\ln(\ell)}{\ln(\varepsilon)}}$. If $\ell = 0$, then $\sigma_u^* = \sigma_u$.

4.3.2 Type-3 Triangular MFs

Now type-3 triangular MFs, $\tilde{\mu}_{\mathbb{A}}(x, u) = $ **ScaleTriScaleGaussIT3MF**, with triangular $FOU(\mathbb{A})$ are used, characterized with upper parameters $[a_1, b_1, c_1]$ for the UMF and λ, ℓ for the LMF to form the $DOU = [\underline{\mu}(x), \overline{\mu}(x)]$. The vertical cuts $\mathbb{A}_{(x)}(u)$ form the $FOU(\mathbb{A})$, these are Gaussian IT2 MFs, $\mu_{\mathbb{A}(x)}(u)$ with parameters $[\sigma_u, m(x)]$ for UMF and LMF λ, ℓ. The IT3 MF $\tilde{\mu}_{\mathbb{A}}(x, u) = $ **ScaleTriScaleGaussIT3MF**(x,{{[a_1, b_1, c_1]},\lambda,[\ell_1, \ell_2]}) is described with the expressions:

$$\overline{\mu}(x) = \begin{cases} 0 & x < a_1 \\ \frac{x-a_1}{b_1-a_1} & a_1 \leq x \leq b_1 \\ \frac{c_1-x}{c_1-b_1} & b_1 < x \leq c_1 \\ 0 & x > c_1 \end{cases} \tag{4.11}$$

The LMF of the DOU, $\mu(x)$, is defined for a_2 and c_2 where these are functions of the UMF parameters (a_1, b_1, c_1) of the DOU, $\overline{\mu}(x)$, and ℓ vector.

$$a_2 = b_1 - (b_1 - a_1)(1 - \ell_1)$$

$$c_2 = b_1 + (c_1 - b_1)(1 - \ell_2)$$

$$\mu(x) = \begin{cases} 0 & x < a_2 \\ \frac{x-a_2}{b_1-a_2} & a_2 \leq x \leq b_1 \\ \frac{c_2-x}{c_2-b_1} & b_1 < x \leq c_2 \\ 0 & x > c_2 \end{cases} \tag{4.12}$$

The function $\mu(x)$ is multiplied by λ to form the LMF of the DOU, $\mu(x)$ is: $\mu(x) = \lambda\mu(x)$. Then $\overline{u}(x)$ and $\underline{u}(x)$ are the DOU upper and lower limits. The range, $\delta(u)$ and radius, σ_u are:

$$\delta(u) = \overline{u}(x) - \underline{u}(x)$$

$$\sigma_u = \frac{\delta(u)}{2\sqrt{3}} + \varepsilon$$

where ε is a machine epsilon number.

The core, $m(x)$, of the IT3 MF $\widetilde{\mu}(x, u)$, is formulated by:

$$m(x) = \begin{cases} 0 & x < a \\ \frac{x-a}{b_1-a} & a \leq x \leq b_1 \\ \frac{c-x}{c-b_1} & b_1 < x \leq c \\ 0 & x > c \end{cases} \tag{4.13}$$

where $a = (a_1 + a_2)/2$ and $c = (c_1 + c_2)/2$. Then, the vertical cuts with IT2 MFs, $\mu_{\mathbb{A}(x)}(u) = [\underline{\mu}_{\mathbb{A}(x)}(u), \overline{\mu}_{\mathbb{A}(x)}(u)]$, are described with the following equations:

$$\overline{\mu}_{\mathbb{A}(x)}(u) = exp\left[-\frac{1}{2}\left(\frac{u - m(x)}{\sigma_u}\right)^2\right] \tag{4.14}$$

$$\underline{\mu}_{\mathbb{A}(x)}(u) = \lambda \cdot exp\left[-\frac{1}{2}\left(\frac{x - m(x)}{\sigma_u^*}\right)^2\right] \tag{4.15}$$

where $\sigma_u^* = \sigma_u\sqrt{\frac{\ln(\ell)}{\ln(\varepsilon)}}, \ell = (\ell_1 + \ell_2)/2$. If $\ell = 0$, then $\sigma_u^* = \sigma_u$.

4.3.3   Inference and Type-Reduction

The format of the Mamdani k-th rule is:

$$R_Z^k : IF x_1 \; is \; \mathbb{F}_1^k \; and \; \ldots and x_i \; is \; \mathbb{F}_i^k \; and \; \ldots \; and \; x_n \; is \; \mathbb{F}_n^k$$

$$THEN \; y_1 \; is \; \mathbb{G}_1^k, \ldots, y_j \; is \; \mathbb{G}_j^k, \ldots, y_m \; is \; \mathbb{G}_m^k$$

where $i = 1,\ldots, n$, $j = 1,\ldots,m$ and $k = 1,\ldots, r$, are the number of inputs, outputs and rules, respectively. To start we represent the rule antecedents as a fuzzy relation A^k, using the Cartesian product with IT3 FSs, $\mathbb{F}_i^k$, and the implication with the consequent of the j-th output, $\mathbb{G}_j^k$; then, the rule $\mathbb{R}_j^k$ is:

$$A^k = \mathbb{F}_1^k \times \cdots \times \mathbb{F}_n^k \tag{4.16}$$

$$\mathbb{R}_j^k = A^k \to \mathbb{G}_j^k \tag{4.17}$$

If $\mathbb{R}_j^k$, is described as a MF of the rules, $\mu_{\mathbb{R}_j^k}(x, y_j)$, then is expressed as:

$$\mu_{\mathbb{R}_j^k}(x, y_j) = \mu_{A^k \to \mathbb{G}_j^k}(x, y_j) \tag{4.18}$$

As a consequence, when the implication is used, $A^k \to \mathbb{G}_j^k$, with multiple antecedents A^k, and consequents $\mathbb{G}_j^k$, connected by the *meet* ($\sqcap$) operator, then

$$\mu_{A^k \to \mathbb{G}_j^k}(x, y_j) = \mu_{\mathbb{F}_1^k \times \cdots \times \mathbb{F}_n^k \to \mathbb{G}_j^k}(x, y_j) = \mu_{\mathbb{F}_1^k \times \cdots \times \mathbb{F}_n^k}(x) \sqcap \mu_{\mathbb{G}_j^k}(y_j)$$

$$\mu_{A^k \to \mathbb{G}_j^k}(x, y_j) = \mu_{\mathbb{F}_1^k}(x_1) \sqcap \cdots \sqcap \mu_{\mathbb{F}_n^k}(x_n) \sqcap \mu_{\mathbb{G}_j^k}(y_j) = \left[\sqcap_{i=1}^n \mu_{\mathbb{F}_i^k}(x_i)\right] \sqcap \mu_{\mathbb{G}_j^k}(y_j)$$

The n-dimensional input, is given by the relation $A_{X\prime}$, as

$$A_{X\prime}(x) = \mu_{X_1}(x_1|x_1\prime) \sqcap \cdots \sqcap \mu_{X_n}(x_n|x_n\prime) = \sqcap_{i=1}^n \mu_{X_i}(x_i|x_i\prime)$$

Each fuzzy relation of $\mathbb{R}_j^k$ determines a fuzzy set of the consequent of the rule $B_j^k = A_{X\prime} \circ \mathbb{R}_j^k$ in **Y** such that

$$\mu_{B_j^k}(y_j|x\prime) = \mu_{A_{X\prime} \circ \mathbb{R}_Z^k}(y_j|x\prime) = \underset{x \in X}{sup}\left[A_{X\prime}(x) \sqcap \mu_{A^k \to \mathbb{G}_j^k}(x, y_j)\right], y \in Y \quad (4.19)$$

We use the *join* ($\sqcup$) operator to aggregate the values $\mu_{B_j^k}(y_j|x\prime)$.

$$\mathbb{B}_j = \mathbb{B}_j^1 \cup \cdots \cup \mathbb{B}_j^k \cup \cdots \cup \mathbb{B}_j^r = \cup_{k=1}^r \mathbb{B}_j^k$$

$$\mu_{\mathbb{B}_j}\left(y_j|x\prime\right) = \mu_{\mathbb{B}_j^1}\left(y_j|x\prime\right) \sqcup \cdots \sqcup \mu_{\mathbb{B}_j^k}\left(y_j|x\prime\right) \sqcup \cdots \sqcup \mu_{\mathbb{B}_j^r}\left(y_j|x\prime\right)$$

$$\mu_{\mathbb{B}_j}\left(y_j|x\prime\right) = \sqcup_{k=1}^r \mu_{\mathbb{B}_j^k}\left(y_j|x\prime\right)$$

$$\mu_{\mathbb{B}_j}\left(y_j|x\prime\right) = \sqcup_{k=1}^r \left[\overset{\sim k}{\Phi}\left(x\prime\right) \sqcup \mu_{\mathbb{G}_j^k}\left(y_j\right) \right]$$

Finally, the type-reduction is postulated as follows:

$$\hat{y}_j = typeReduction\left(y_j, \mu_{\mathbb{B}_j}\left(y_j|x\prime\right)\right) \tag{4.20}$$

and this produces the final result.

4.4 Fuzzy Fractal Prediction

Let $z_1, z_2, \ldots, z_n$ be a time series. If a clustering algorithm offers n clusters $C_1, C_2, \ldots, C_n$, then a fuzzy system can be outlined. Cluster complexity can be estimated by the dimensions, linear is dim_1 and non-linear dim_2, with values $x_1, x_2, \ldots, x_n$, and $y_1, y_2, \ldots, y_n$, respectively. Then, the fuzzy system is formulated as:

$$\begin{array}{l} \text{If dim}_1 \text{ is } x_1 \text{ and dim}_2 \text{ is } y_1 \text{ then prediction is } C_1 \\ \text{If dim}_1 \text{ is } x_2 \text{ and dim}_2 \text{ is } y_2 \text{ then prediction is } C_2 \\ \cdots \\ \text{If dim}_1 \text{ is } x_n \text{ and dim}_2 \text{ is } y_n \text{ then prediction is } C_n \end{array} \tag{4.21}$$

For applying (4.21), MFs for the FDs need to be defined. The four input variables are: the linear FD of confirmed (LFDC), nonlinear FD of confirmed (NLFDC), linear FD of death (LFDD), and nonlinear FD of death (NLFDD) cases. Low (L) and high (H) are the FD values. The output is the Prediction Increment (ΔP) with 3 values: High (H), Medium (M) and Low (L). The total process is depicted in Fig. 4.2. Now ΔP is calculated with Eq. (4.20). Finally, ΔP is aggregated to the previous value, to estimate the next prediction, which is P_{n+1}. The method is delineated in Fig. 4.3.

The algorithm is presented in Fig. 4.4.

The rules were constructed with problem knowledge. The rules are listed in Fig. 4.5. The output MFs are exhibited in Fig. 4.6 (triangular and trapezoidal). In Fig. 4.7 the input MFs are depicted. Two Gaussian MFs are utilized for both values.

In Table 4.1 we list the particular MF parameters, which were found by experimentation. Basically, Table 4.1 summarizes the center (c) and deviation (σ) of the MFs.

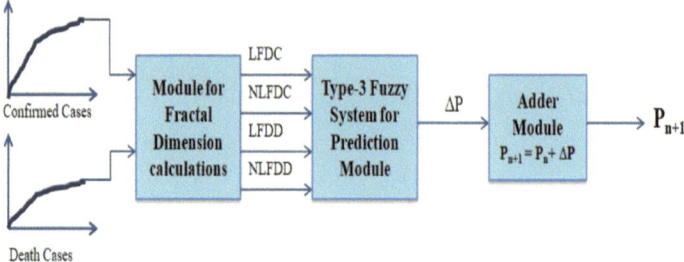

Fig. 4.2 Fuzzy fractal method

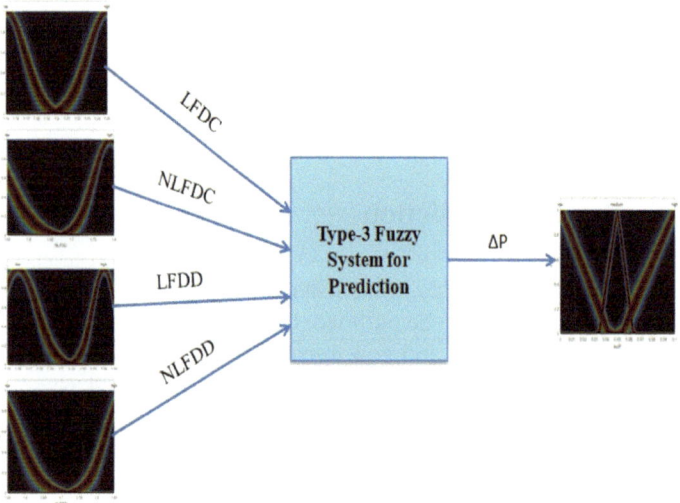

Fig. 4.3 Fuzzy fractal model for prediction

Fig. 4.4 Algorithm of prediction

Algorithm: IT3FS+FD
Data: z_1, z_2, ...z_n
Goal: Predict z_{n+1}, z_{n+2}, ...z_{n+m}
Establish Parameters: Triangular: a1, b1, c1, λ, ℓ. Gaussian: σ, m, λ, ℓ.
  For *i= 1:m*
    Estimate FDs with Eq. 4.2
    Estimate ΔPs with Eq. 4.20
    Estimate $P_{n+i} = P_{n+i-1} + \Delta P$
    Estimate Error: $e_{n+i} = z_{n+i} - P_{n+i}$
  End For

In Table 4.2 we exhibit the particular output MF parameters, obtained by experimentation, for the triangular MFs.

Fig. 4.5 Rules of the forecasting knowledge

1. If (LFDC is low) and (NLFDC is high) and (LFDD is low) and (NLFDD is low) then (IncP is High)(1)
2. If (LFDC is low) and (NLFDC is low) and (LFDD is low) and (NLFDD is low) then (IncP is Medium)(1)
3. If (LFDC is low) and (NLFDC is low) and (LFDD is low) and (NLFDD is high) then (IncP is Low)(1)
4. If (LFDC is high) and (NLFDC is low) and (LFDD is low) and (NLFDD is high) then (IncP is High)(1)
5. If (LFDC is high) and (NLFDC is high) and (LFDD is high) and (NLFDD is high) then (IncP is High)(1)
6. If (LFDC is low) and (NLFDC is high) and (LFDD is low) and (NLFDD is high) then (IncP is High)(1)

Fig. 4.6 Output MFs of the model

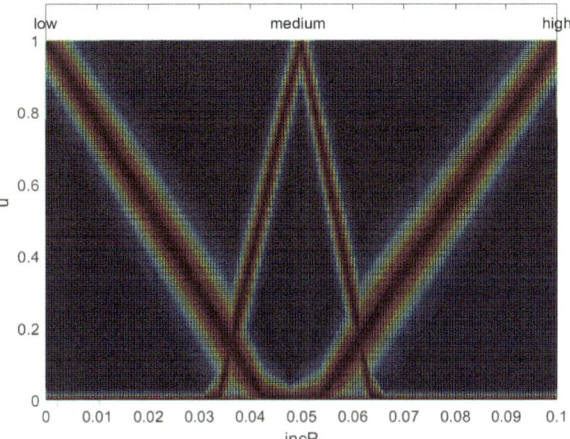

Fig. 4.7 MFs for the LFDD variable

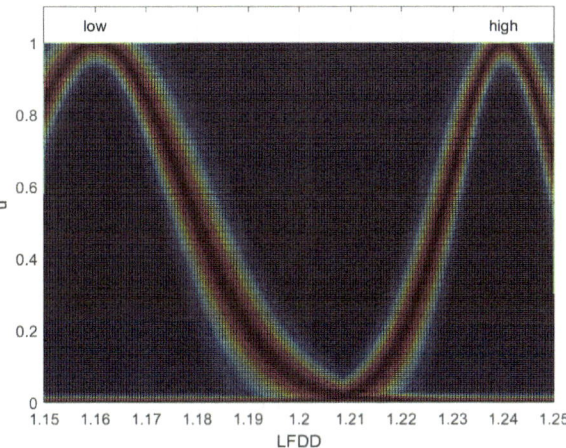

Table 4.1 Parameters utilized in the inputs (L = low, High = H)

Variable	MFs	σ	m
LFDC	L	0.022	1.150
LFDC	H	0.024	1.250
NLFDC	L	0.076	1.490
NLFDC	H	0.049	1.790
LFDD	L	0.021	1.160
LFDD	H	0.013	1.240
NLFDD	L	0.083	1.510
NLFDD	H	0.066	1.880

Table 4.2 Parameters utilized in the outputs

Variable	MFs	a	B	c
ΔP	L	0.000	0.000	0.050
ΔP	M	0.030	0.049	0.065
ΔP	H	0.045	0.100	0.100

In Fig. 4.8 we exhibit a 3-D view of the type-3 MF, where we appreciate the upper and lower MFs.

In Fig. 4.9 we exhibit a different view of the MF.

In Fig. 4.10 the surface representing the model can be found.

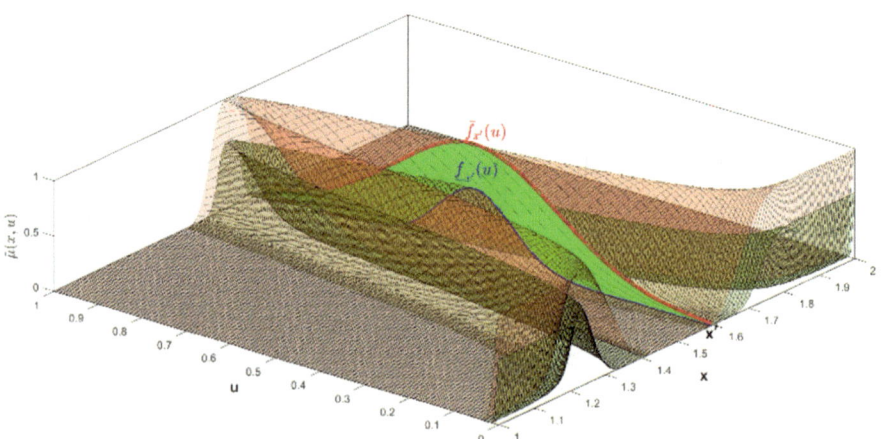

Fig. 4.8 IT3 FS with IT3MF $\tilde{\mu}(x, u)$ where $\underline{\mu}(x, u)$ is the LMF and $\overline{\mu}(x, u)$ is the UMF

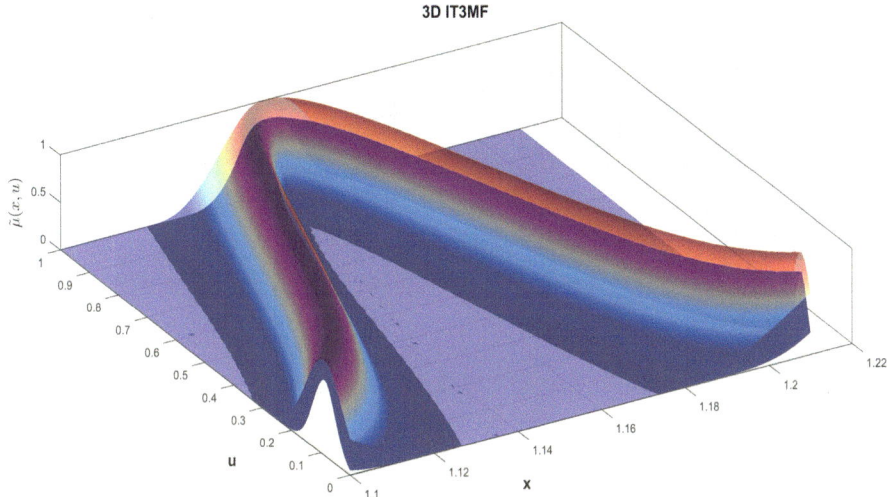

Fig. 4.9 MF of the IT3FS exhibiting the volume in 3D

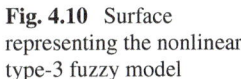

Fig. 4.10 Surface representing the nonlinear type-3 fuzzy model

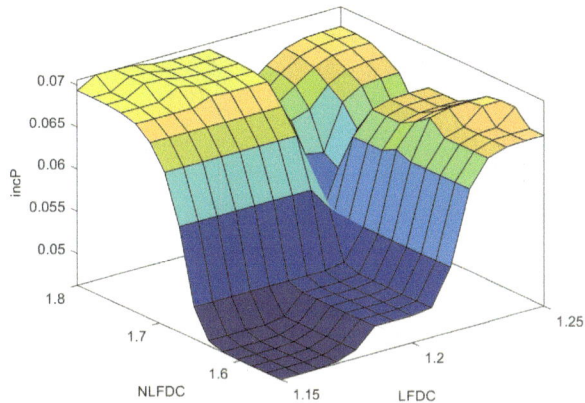

4.5 Simulation Results

The experiments were performed with a dataset used from the Humanitarian Data Exchange (HDX) [8], which includes COVID-19 data from countries, where cases have occurred from January 22, 2020 to April 15, 2020. The reason for using this time interval is to compare results with previous works using type-1 fuzzy logic [37]. The consulted datasets are: time_series_covid19_confirmed_global, time_series_covid19_recovered_global, and time_series_covid19_deaths_global. The data consists of confirmed, recovered and deaths cases. Figure 4.11 shows a plot for Belgium with cases for the period. In Fig. 4.12 a similar plot for Italy is illustrated.

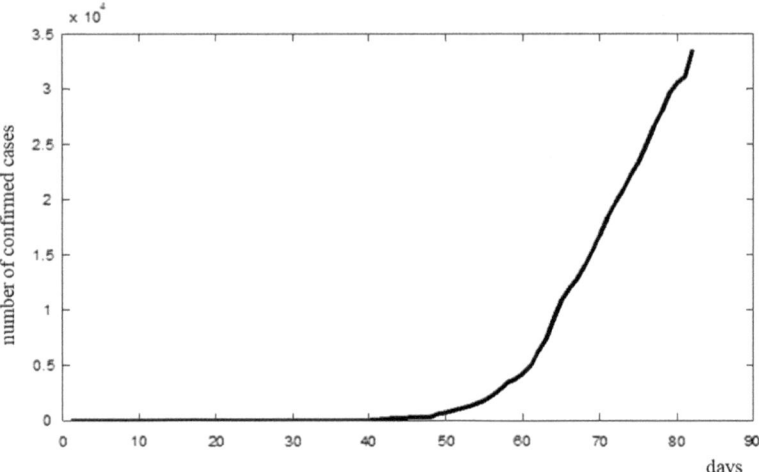

Fig. 4.11 Plot of Belgium confirmed cases

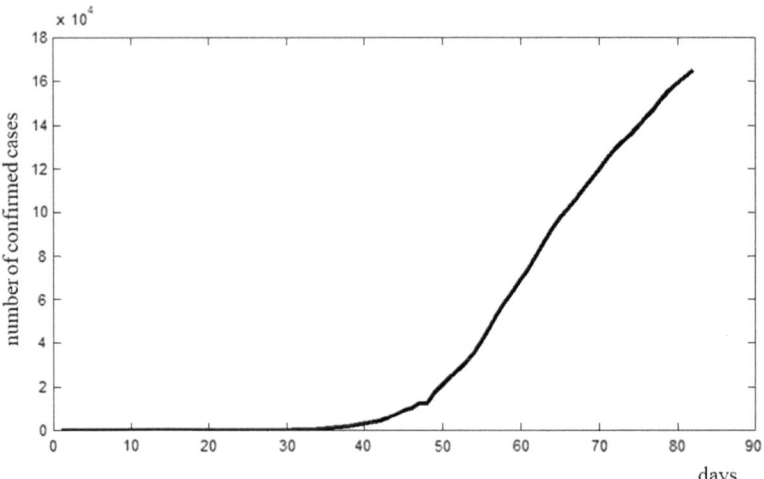

Fig. 4.12 Plot of Italy confirmed cases

Figure 4.13 shows a plot of Belgium COVID-19 deaths, for the period. In Fig. 4.14 we exhibit in a similar way the trend for Italy.

Based on the data, FD values are approximations and reported in Table 4.3. The resulting increments (ΔP) are highlighted in bold.

Figure 4.15 depicts the Belgium prediction, where predictions are close to real data. Figure 4.16 reports the Germany prediction. Figure 4.17 reports the USA prediction. Lastly, in Figs. 4.18 and 4.19 we exhibit the predictions for Spain and Italy, respectively.

Fig. 4.13 Plot of Belgium death cases

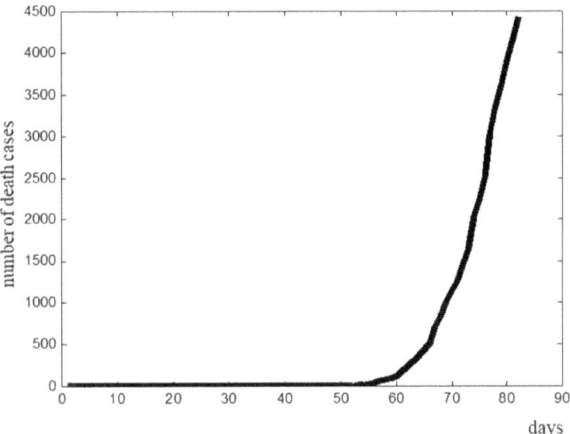

Fig. 4.14 Plot of Italy death cases

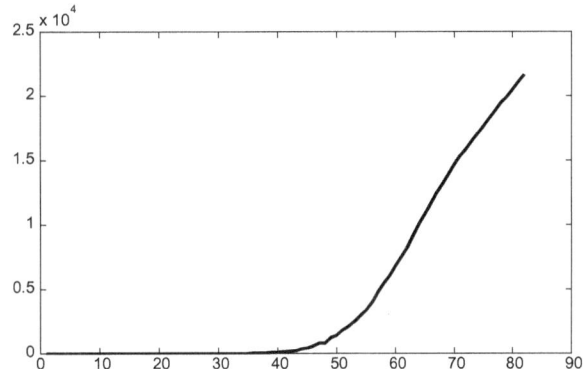

Table 4.3 Dimensions and increments produced by the method

Metric	FDs								
	Belgium (B)	France (F)	Germany (G)	Iran (Ir)	Italy(I)	Spain(S)	Turkey (T)	UK	US
LFDC	1.186	1.190	1.202	1.191	1.194	1.186	1.204	1.207	1.204
NLFDC	1.748	1.744	1.615	1.721	1.722	1.775	1.608	1.624	1.593
LFDD	1.208	1.190	1.178	1.204	1.189	1.181	1.202	1.212	1.187
NLFDD	1.604	1.788	1.710	1.623	1.614	1.789	1.596	1.601	1.804
AP	**0.069**	**0.076**	**0.049**	**0.071**	**0.073**	**0.077**	**0.050**	**0.050**	**0.067**

Table 4.4 reports forecasts for 9 countries and these are illustrated in Fig. 4.19. To model was built with COVID-19 cases from January 22 to April 15 of 2020 were used. The forecast is for 10 days (April 16 to 25 of 2020).

In Table 4.5 a comparative of prediction errors is presented.

Fig. 4.15 Prediction of
Belgium confirmed cases

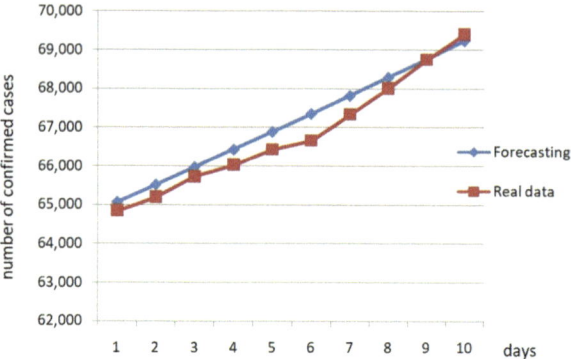

Fig. 4.16 Prediction for
Germany

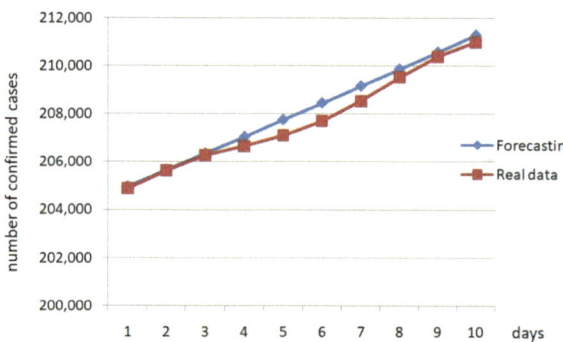

Fig. 4.17 Prediction of
USA confirmed cases

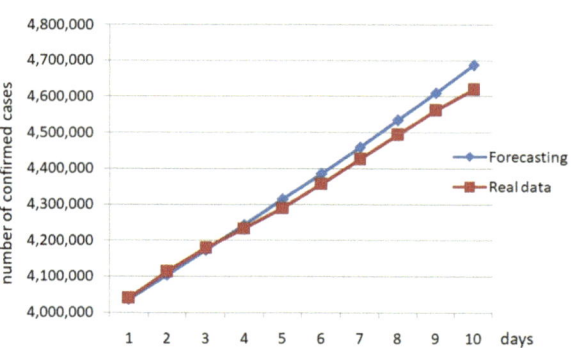

We optimized λ and ℓ of the MFs to corroborate if better results are achieved. As an initial test, we decided to use the Firefly Algorithm, as in [38], because it is relatively easy to use.

We performed a comparison of type-3 prediction versus type-2. Now we considered in this experiment 12 countries to be able to compare with the work in [38]. Table 4.6 depicts a comparison for COVID-19 prediction errors for 12 countries in

Fig. 4.18 Prediction for Spain

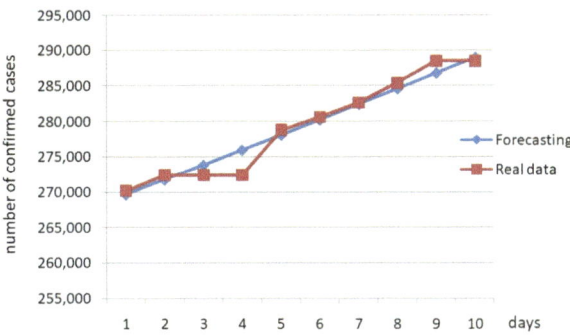

Fig. 4.19 Forecast of Italy cases

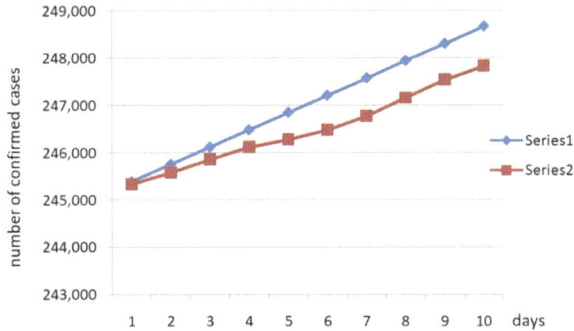

which type-3 is better in 11 of the 12 cases, demonstrating that optimization improved the results.

We have also previously compared the same approach using type-2 with respect to type-1 fuzzy logic in a previous paper, showing that type-2 was better [38], so in this way we can conclude that type-3 fuzzy outperforms both type-2 and type-1 in COVID-19 prediction.

4.6 Summary

A hybrid approach for prediction was proposed. In addition, the approach was also compared with the previous work [37] for the same period of time showing that forecasting errors are lower with the utilization of interval type-3 fuzzy logic. Future work may consist on applying the proposed approach on similar problems [39–41], as well as generalizing the use of fuzzy logic to general type-3 and considering granularity concepts [42–46], which we envision will enable a better representation of the uncertainty in the prediction process. Finally, we could perform a hybridization with other prediction methods [47–50] to test if it is possible to enhance prediction accuracy.

Table 4.4 Confirmed cases prediction

B	F	G	Ir	I	S	T	UK	USA
6.507×10^4	2.053×10^5	2.049×10^5	2.836×10^5	2.453×10^5	2.696×10^5	2.234×10^5	2.972×10^5	4.036×10^6
6.552×10^4	2.060×10^5	2.056×10^5	2.858×10^5	2.457×10^5	2.717×10^5	2.244×10^5	2.980×10^5	4.104×10^6
6.597×10^4	2.068×10^5	2.063×10^5	2.881×10^5	2.461×10^5	2.738×10^5	2.254×10^5	2.989×10^5	4.173×10^6
6.643×10^4	2.075×10^5	2.070×10^5	2.904×10^5	2.464×10^5	2.759×10^5	2.264×10^5	2.997×10^5	4.243×10^6
6.689×10^4	2.082×10^5	2.077×10^5	2.927×10^5	2.468×10^5	2.780×10^5	2.274×10^5	3.006×10^5	4.314×10^6
6.735×10^4	2.090×10^5	2.084×10^5	2.950×10^5	2.472×10^5	2.802×10^5	2.284×10^5	3.015×10^5	4.386×10^6
6.782×10^4	2.097×10^5	2.091×10^5	2.973×10^5	2.475×10^5	2.823×10^5	2.295×10^5	3.023×10^5	4.459×10^6
6.829×10^4	2.104×10^5	2.098×10^5	2.997×10^5	2.479×10^5	2.845×10^5	2.305×10^5	3.032×10^5	4.534×10^6
6.876×10^4	2.112×10^5	2.105×10^5	3.020×10^5	2.483×10^5	2.867×10^5	2.315×10^5	3.041×10^5	4.610×10^6
6.924×10^4	2.119×10^5	2.113×10^5	3.044×10^5	2.486×10^5	2.889×10^5	2.326×10^5	3.050×10^5	4.688×10^6

Table 4.5 Comparison of forecasting errors for the 9 countries between the proposed approach with type-3 fuzzy logic and type-1 fuzzy of previous work [37]

	B	F	G	Ir	I	S	T	UK	USA
Type-1	0.00613	0.00381	0.00300	0.02741	0.00046	0.00277	0.00029	0.01837	0.01438
Type-3	0.00231	0.00498	0.00143	0.00739	0.00341	0.00164	0.00330	0.00348	0.01462

Table 4.6 Comparison of predictions for type-3 versus type-2 in 12 countries based on Absolute Errors

C	Comparison					
	Type-2 [38]			Type-3 (This chapter)		
	Best	Avg	Worst	Best	Avg	Worst
Brazil (B)	0.0000030	**0.01970**	0.14200	0.0000018	0.02460	0.1060
China (Ch)	0.0018400	0.06980	0.29600	0.0005200	**0.02970**	0.1610
F	0.0000060	0.02060	0.19400	0.0000041	**0.00710**	0.0606
G	0.0008020	0.08550	0.41100	0.0000409	**0.03020**	0.1010
India (In)	0.0000001	0.00880	0.15400	0.00000030	**0.00300**	0.0205
Ir	0.0000112	0.01780	0.09820	0.00000075	**0.01430**	0.1050
I	0.0000075	0.04540	0.29200	0.0000124	**0.01760**	0.0832
M	0.0000286	0.009140	0.18100	0.0000148	**0.001490**	0.0300
Poland (P)	0.0000831	0.020500	0.42800	0.000057	**0.006080**	0.0585
S	0.0005820	0.000917	0.00154	0.000556	**0.000837**	0.0015
UK	0.0000280	0.007070	0.16200	0.000269	**0.012600**	0.0998
USA	0.00000315	0.008020	0.06040	0.00000062	**0.005320**	0.0949

References

1. B. Mandelbrot, *The Fractal Geometry of Nature* (W.H. Freeman and Company, 1987)
2. O. Castillo, P. Melin, A new method for fuzzy estimation of the fractal dimension and its applications to time series analysis and pattern recognition, in *Proceedings of NAFIPS'2000, Atlanta, GA, USA*, pp. 451–455
3. R. Yager, D. Filev, Generation of fuzzy rules by mountain clustering. Intell. Fuzzy Syst. **2**(3), 209–219 (1994)
4. L.A. Zadeh, The concept of a linguistic variable and its application to approximate reasoning. Inf. Sci. **8**, 43–80 (1975)
5. M. Sugeno, G.T. Kang, Structure identification of fuzzy model. Fuzzy Sets Syst. **28**, 15–33 (1988)
6. J.R. Jang, C.T. Sun, E. Mizutani, Neuro-Fuzzy and Soft Computing (Prentice Hall, 1997)
7. P. Melin, O. Castillo, An adaptive model-based neuro-fuzzy-fractal controller for biochemical reactors in the food industry, in *Proceedings of IJCNN'98, IEEE Computer Society Press, Alaska, USA*, vol. 1 (1998), pp. 106–111.
8. The Humanitarian Data Exchange (HDX). https://data.humdata.org/dataset/novel-coronavirus-2019-ncov-cases. Accessed 31 Mar 2020
9. M.A. Shereen, S. Khan, A. Kazmi, N. Bashir, R. Siddique, COVID-19 infection: origin, transmission, and characteristics of human coronaviruses. J. Adv. Res. **24**, 91–98 (2020)
10. C. Sohrabi, Z. Alsafi, N. O'Neill, M. Khan, A. Kerwan, A. Al-Jabir, C. Iosifidis, R. Agha, World health organization declares global emergency: a review of the 2019 novel coronavirus (COVID-19). Int. J. Surg. **76**, 71–76 (2020)
11. I.D. Apostolopoulos, T. Bessiana, Covid-19: automatic detection from X-Ray images utilizing transfer learning with convolutional neural networks (2020). arXiv:2003.11617
12. S.A. Sarkodie, P.A. Owusu, Investigating the Cases of Novel Coronavirus Disease (COVID-19) in China Using dynamic statistical techniques (2020). Available at SSRN 3559456

13. B.R. Beck, B. Shin, Y. Choi, S. Park, K. Kang, Predicting commercially available antiviral drugs that may act on the novel coronavirus (SARS-CoV-2) through a drug-target interaction deep learning model. Comput. Struct. Biotechnol. J. **18**, 784–790 (2020)
14. L. Zhong, L. Mu, J. Li, J. Wang, Z. Yin, D. Liu, Early prediction of the 2019 novel coronavirus outbreak in the Mainland China based on simple mathematical model. IEEE Access **8**, 51761–51769 (2020)
15. M.N. Kamel Boulos, E.M. Geraghty, Geographical tracking and mapping of coronavirus disease COVID-19/severe acute respiratory syndrome coronavirus 2 (SARS-CoV-2) epidemic and associated events around the world: how 21st century GIS technologies are supporting the global fight against outbreaks and epidemics. Int. J. Health Geogr. **19**, 8 (2020). https://doi.org/10.1186/s12942-020-00202-8
16. P. Gao, H. Zhang, Z. Wu, J. Wang, Visualising the expansion and spread of coronavirus disease 2019 by cartograms. Environ. Plan A (2020). https://doi.org/10.1177/0308518X20910162
17. A.S.R.S. Rao, J.A. Vazquez, Identification of COVID-19 can be quicker through artificial intelligence framework using a mobile phone-based survey in the populations when Cities/Towns are under quarantine. Infect. Control Hosp. Epidemiol. (2020). https://doi.org/10.1017/ice.2020.61
18. K.C. Santosh, AI-driven tools for coronavirus outbreak: need of active learning and cross-population Train/Test models on Multitudinal/Multimodal data. J. Med. Syst. **44**(5). https://doi.org/10.1007/s10916-020-01562-1
19. B. Robson, Computers and viral diseases. Preliminary bioinformatics studies on the design of a synthetic vaccine and a preventative peptidomimetic antagonist against the SARS-CoV-2 (2019-nCoV, COVID-19) coronavirus. Comput. Biol. Med. **119**, 1–19 (2020)
20. D. Fanelli, F. Piazza, Analysis and forecast of COVID-19 spreading in China, Italy and France. Chaos Solitons Fractals **134**, 1–5 (2020)
21. S. Contreras et al., A multi-group SEIRA model for the spread of COVID-19 among heterogeneous populations. Chaos Solitons Fractals **136**, 1099325 (2020)
22. N. Crokidakis, COVID-19 spreading in Rio de Janeiro, Brazil: do the policies of social isolation really work? Chaos Solitons Fractals **136**, 109930 (2020)
23. M.S. Adbo et al., On a comprehensive model of the novel coronavirus (COVID-19) under Mittag-Leffler derivative. Chaos Solitons Fractals **135**, 109867 (2020)
24. S. Boccaletti et al., Modeling and forecasting of epidemic spreading: the case of Covid-19 and beyond. Chaos Solitons Fractals **135**, 109794 (2020)
25. T. Chakraborty, I. Ghosh, Real-time forecasts and risk assessment of novel coronavirus (COVID-19) cases: a data-driven analysis. Chaos Solitons Fractals **135**, 109850 (2020)
26. M. Mandal et al., A model based study on the dynamics of COVID-19: prediction and control. Chaos Solitons Fractals **136**, 109889 (2020)
27. P. Melin, J.C. Monica, D. Sanchez, O. Castillo, Analysis of spatial spread relationships of coronavirus (COVID-19) pandemic in the world using self organizing maps. Chaos Solitons Fractals **138**(109917), 1–7 (2020)
28. P. Melin, J.C. Monica, D. Sanchez, O. Castillo, Multiple ensemble neural network models with fuzzy response aggregation for predicting COVID-19 time series: the case of Mexico. Healthcare **8**, 181 (2020)
29. O. Castillo, P. Melin, Developing a new method for the identification of microorganisms for the food industry using the fractal dimension. J. Fractals **2**(3), 457–460 (1994)
30. O. Castillo, P. Melin, A new fuzzy inference system for reasoning with multiple differential equations for modelling complex dynamical systems, in *Proceedings of CIMCA 1999, IOS Press, Vienna Austria* (1999), pp.224–229
31. J.C. Bezdek, *Pattern Recognition with Fuzzy Objective Function Algorithms* (Plenum Press, 1981)
32. O. Castillo, P. Melin, A new fuzzy-fractal-genetic method for automated mathematical modelling and simulation of robotic dynamic systems, in *Proceedings of FUZZ'98, IEEE Press, Alaska, USA*, vol. 2 (1998), pp. 1182–1187
33. L. Zadeh, Fuzzy sets. Inf. Control **8** (1965)

34. A. Mohammadzadeh, M.H. Sabzalian, W. Zhang, An interval type-3 fuzzy system and a new online fractional-order learning algorithm: theory and practice. IEEE Trans. Fuzzy Syst. **28**(9), 1940–1950 (2020)
35. J.T. Rickard, J. Aisbett, G. Gibbon, Fuzzy subsethood for fuzzy sets of type-2 and generalized type-n. IEEE Trans. Fuzzy Syst. **17**(1), 50–60 (2009)
36. O. Castillo, Towards finding the optimal n in designing type-n fuzzy systems for particular classes of problems: a review. Appl. Comput. Math. **17**(1), 3–9 (2018)
37. O. Castillo, P. Melin, Forecasting of COVID-19 time series for countries in the world based on a hybrid approach combining the fractal dimension and fuzzy logic. Chaos Solitons Fractals **140**, 110242 (2020)
38. P. Melin, D. Sánchez, J.C. Monica, O. Castillo, Optimization using the firefly algorithm of ensemble neural networks with type-2 fuzzy integration for COVID-19 time series prediction. Soft. Comput. **1**, 1–38 (2021)
39. E. Ontiveros-Robles, P. Melin, O. Castillo, Comparative analysis of noise robustness of type 2 fuzzy logic controllers. Kybernetika **54**(1), 175–201 (2018)
40. O. Torrealba-Rodriguez, R.A. Conde-Gutiérrez, A.L. Hernández-Javier, Modeling and prediction of COVID-19 in Mexico applying mathematical and computational models. Chaos Solitons Fractals **138**, 1–8 (2020)
41. T. Sun, Y. Wang, Modeling COVID-19 epidemic in Heilongjiang province, China. Chaos Solitons Fractals **138**, 1–5 (2020)
42. O. Castillo, *Type-2 Fuzzy Logic in Intelligent Control Applications*. (Springer, Berlin, 2012)
43. M.A. Sanchez, O. Castillo, J.R. Castro, P. Melin, Fuzzy granular gravitational clustering algorithm for multivariate data. Inf. Sci. **279**, 498–511 (2014)
44. C.I. González, P. Melin, J.R. Castro, O. Mendoza, O. Castillo, An improved Sobel edge detection method based on generalized type-2 fuzzy logic. Soft. Comput. **20**(2), 773–784 (2016)
45. E. Ontiveros, P. Melin, O. Castillo, High order α-planes integration: a new approach to computational cost reduction of general type-2 fuzzy systems. Eng. Appl. AI **74**, 186–197 (2018)
46. A. Mohammadzadeh, O. Castillo, S.S. Band et al., A novel fractional-order multiple-model type-3 fuzzy control for nonlinear systems with unmodeled dynamics. Int. J. Fuzzy Syst. **23**, 1633–1651 (2021)
47. O. Castillo, J.R. Castro, P. Melin, Interval type-3 fuzzy aggregation of neural networks for multiple time series prediction: the case of financial forecasting. Axioms **11**, 251 (2022). https://doi.org/10.3390/axioms11060251
48. M. Ramirez, P. Melin, A new perspective for multivariate time series decision making through a nested computational approach using type-2 fuzzy integration. Axioms **12**, 385 (2023). https://doi.org/10.3390/axioms12040385
49. M. Ramírez, P. Melin, O. Castillo, Interval type-3 fuzzy aggregation for hybrid-hierarchical neural classification and prediction models in decision-making. Axioms **12**, 906 (2023). https://doi.org/10.3390/axioms12100906
50. P. Melin, D. Sánchez, J.R. Castro, O. Castillo, Design of type-3 fuzzy systems and ensemble neural networks for COVID-19 time series prediction using a firefly algorithm. Axioms **11**, 410 (2022). https://doi.org/10.3390/axioms11080410

Chapter 5
Type-3 Fuzzy Aggregation of Neural Networks

Abstract In this chapter we are presenting the aggregation of NNs for prediction. Type-3 aggregation is employed for improving the prediction.

Keywords Type-3 fuzzy logic · Aggregation · Time series

5.1 Introduction

It is known in that using type-1 fuzzy logic helps to enhance results in many problems [1, 2]. Later type-1 evolved to type-2 fuzzy systems [3, 4]. Initially, interval type-2 was applied to several problems in different areas [5, 6]. Results show that interval type-2 surpasses type-1 in situations with higher noise levels [6–8]. Later, general type-2 fuzzy systems were considered to manage higher uncertainty levels with striking results, like in [9–11]. Recently, it is now a reality that type-3 can solve more complex problems [12–14].

Recently, the rapid COVID-19 propagation has been recognized, spreading worldwide. In particular, in Europe several countries were affected by COVID-19 [15–20]. In case of the American continent also a significant number of cases occurred due COVID-19 [21–23]. There are also works on COVID-19 prediction behavior in space and time [24–26]. Additionally, it was noted that waves of COVID-19 impacted the economy, and for this reason both time series are considered. The contribution is the type-3 fuzzy aggregation of NNs, as is not published before.

The structure of this chapter: Sect. 5.2 highlights terminology of type-3, Sect. 5.3 describes the prediction method, Sect. 5.4 lists the results, and Sect. 5.5 provides the conclusions.

5.2 Type-3 Fuzzy Logic

The terminology of type-3 is outlined in this subsection.

© The Author(s), under exclusive license to Springer Nature Switzerland AG 2024   49
O. Castillo and P. Melin, *Type-3 Fuzzy Logic in Time Series Prediction*,
SpringerBriefs in Computational Intelligence,
https://doi.org/10.1007/978-3-031-59714-5_5

Definition 5.1 A type-3 fuzzy set (T3 FS) [27–30], denoted by $A^{(3)}$, is represented by a function, called MF of $A^{(3)}$, in the Cartesian product $X \times [0, 1] \times [0, 1]$ in $[0, 1]$, where X is the universe of the primary variable of $A^{(3)}$, x. The MF of $\mu_{A^{(3)}}$ is denoted by $\mu_{A^{(3)}}(x, u, v)$ (or $\mu_{A^{(3)}}$) and it is called a T3 MF of the T3 FS,

$$\mu_{A^{(3)}} : X \times [0, 1] \times [0, 1] \rightarrow [0, 1]$$
$$A^{(3)} = \{(x, u(x), v(x, u), \mu_{A^{(3)}}(x, u, v)) | x \in X, u \in U \subseteq [0, 1], v \in V \subseteq [0, 1]\}$$
$$(5.1)$$

where U is the universe for the secondary variable u and V is the universe for tertiary variable v. If the tertiary MF is uniformly equal to 1 then an IT3 FS is produced.

Figure 5.1 shows and IT3 FS with IT3MF $\tilde{\mu}(x, u)$, where $\underline{\mu}(x, u)$ is the LMF and $\overline{\mu}(x, u)$ is the UMF. The embedded secondary T1 MFs in x/ of $\underline{A}$ and $\overline{A}$ are $\underline{f}_{x/}(u)$ and $\overline{f}_{x/}(u)$.

This function can be represented as, $\tilde{\mu}_{\mathbb{A}}(x, u) =$ ScaleGaussScaleGauss IT3MF, with Gaussian $FOU(\mathbb{A})$, with upper parameters $[\sigma, m]$ for the UMF and λ, ℓ for the LMF to form the $DOU = [\underline{\mu}(x), \overline{\mu}(x)]$. The vertical cuts $\mathbb{A}_{(x)}(u)$ form the $FOU(\mathbb{A})$, and are Gaussian IT2 MFs, $\mu_{\mathbb{A}(x)}(u)$ with parameters $[\sigma_u, \mathrm{m}(x)]$ for UMF and LMF with λ, and ℓ. The IT3 MF, $\tilde{\mu}_{\mathbb{A}}(x, u) =$ ScaleGaussScaleGaussIT3MF(x,$\{\{[\sigma, m]\}, \lambda, \ell\}$) is postulated with expressions:

$$\overline{u}(x) = exp\left[-\frac{1}{2}\left(\frac{x - m}{\sigma}\right)^2 \right] \tag{5.2}$$

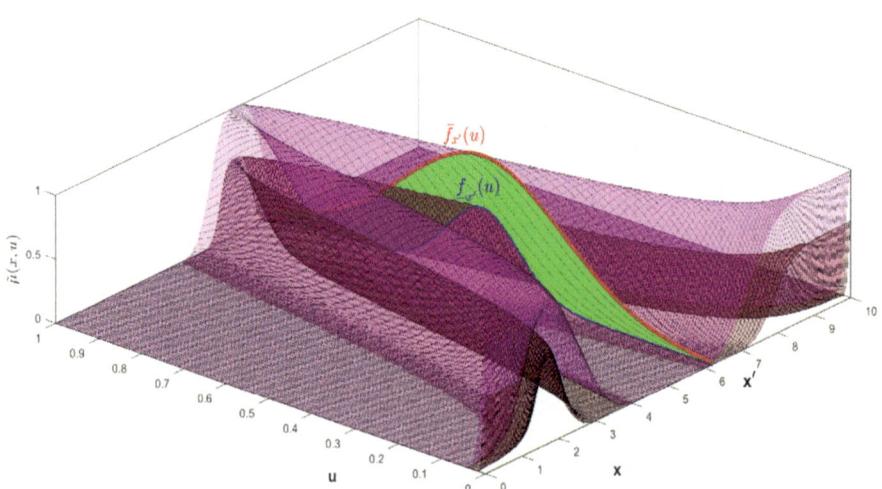

Fig. 5.1 Fuzzy set with $\tilde{\mu}(x, u)$

$$\underline{u}(x) = \lambda \cdot exp\left[-\frac{1}{2}\left(\frac{x-m}{\sigma^*}\right)^2\right] \tag{5.3}$$

where $\sigma^* = \sigma\sqrt{\frac{\ln(\ell)}{\ln(\varepsilon)}}$, ε is the machine epsilon. If $\ell = 0$, then $\sigma^* = \sigma$. Then $\overline{u}(x)$ and $\underline{u}(x)$ are the DOU upper and lower limits. The range, $\delta(u)$ and radius, σ_u are:

$$\delta(u) = \overline{u}(x) - \underline{u}(x) \tag{5.4}$$

$$\sigma_u = \frac{\delta(u)}{2\sqrt{3}} + \varepsilon \tag{5.5}$$

The apex or core, $m(x)$, of the IT3 MF $\tilde{\mu}(x, u)$, is postulated by:

$$m(x) = exp\left[-\frac{1}{2}\left(\frac{x-m}{\rho}\right)^2\right] \tag{5.6}$$

where $\rho = (\sigma + \sigma^*)/2$. Then, the vertical cuts with IT2 MF, $\mu_{\mathbb{A}(x)}(u) = [\underline{\mu}_{\mathbb{A}(x)}(u), \overline{\mu}_{\mathbb{A}(x)}(u)]$, are described by equations:

$$\overline{\mu}_{\mathbb{A}(x)}(u) = exp\left[-\frac{1}{2}\left(\frac{u - u(x)}{\sigma_u}\right)^2\right] \tag{5.7}$$

$$\underline{\mu}_{\mathbb{A}(x)}(u) = \lambda \cdot exp\left[-\frac{1}{2}\left(\frac{u - u(x)}{\sigma_u^*}\right)^2\right] \tag{5.8}$$

where $\sigma_u^* = \sigma_u\sqrt{\frac{\ln(\ell)}{\ln(\varepsilon)}}$. If $\ell = 0$, then $\sigma_u^* = \sigma_u$.

5.3 Method

Figure 5.2 exhibits the method architecture, where we notice that two time series enter the modules (NN$_1$ and NN$_2$) and predictions P$_1$ and P$_2$ are estimated with increments ΔP$_1$ and ΔP$_2$. These increments are the inputs for aggregation that will obtain an increment for series number 1 and the prediction is produced.

The fuzzy rules for aggregation are:

1. If (ΔP$_1$ is H) and (ΔP$_2$ is L), then (ΔP is P).
2. If (ΔP$_1$ is H) and (ΔP$_2$ is M), then (ΔP is NS).
3. If (ΔP$_1$ is H) and (ΔP$_2$ is H), then (ΔP is NL).
4. If (ΔP$_1$ is M) and (ΔP$_2$ is L), then (ΔP is P).
5. If (ΔP$_1$ is M) and (ΔP$_2$ is M), then (ΔP is NS).

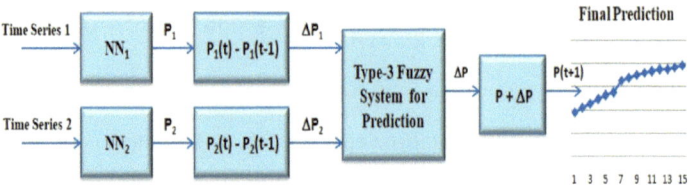

Fig. 5.2 Proposed ensemble with type-3 fuzzy aggregation

6. If (ΔP_1 is M) and (ΔP_2 is H), then (ΔP is NL).
7. If (ΔP_1 is L) and (ΔP_2 is L), then (ΔP is P).
8. If (ΔP_1 is L) and (ΔP_2 is M), then (ΔP is NS).
9. If (ΔP_1 is L) and (ΔP_2 is H), then (ΔP is NL).

Where H is high, M is medium, L is low, P is positive, NL is negative large and NS is negative small. The rules are outlined based on knowledge. This knowledge was employed in outlining the rules. The system (Fig. 5.3) has as inputs the increment on prediction of each NN, P_1 and P_2, respectively. After the defuzzification, the output is the increment of series number 1. In particular, as the first time series the Dow Jones (DJ) index is used, and as the second one COVID-19 cases are used, and considering their relation the Dow Jones prediction is enhanced.

In Table 5.1 we show the particular MF parameters, obtained by trial and error. Table 5.1 exhibits the centers and deviations of the MFs.

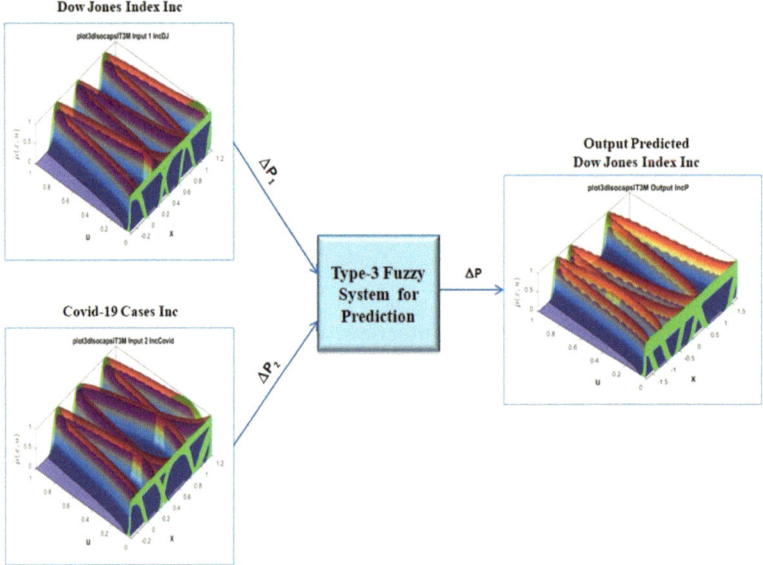

Fig. 5.3 System to calculate the weights

Table 5.1 Parameters for the MFs (Small = S, medium = M, high = H)

Variable	MF	σ	m
ΔP_1	S	0.127	0.0
ΔP_1	M	0.130	0.5
ΔP_1	H	0.250	1.0
ΔP_2	S	0.200	0.0
ΔP_2	M	0.150	0.5
ΔP_2	H	0.300	1.0
ΔP	S	0.150	−1.0
ΔP	M	0.180	−0.5
ΔP	H	0.250	1.0

Regarding the λ and ℓ parameters, after experimentation, they were found to be 0.9 and 0.2 for both inputs. However, for both outputs, they were found to be 0.8 and 0.2.

Figures 5.4 and 5.5 exhibit input MFs for both errors, respectively. Figure 5.6 exhibits the output MFs. In Fig. 5.7 we depict a view of the surface with the effect of w_1 on the errors e_1 and e_2.

In Fig. 5.8 we exhibit the inference for a specific value, and then in Fig. 5.9 the type-reduction and defuzzification.

Fig. 5.4 MFs of increment of DJ variable

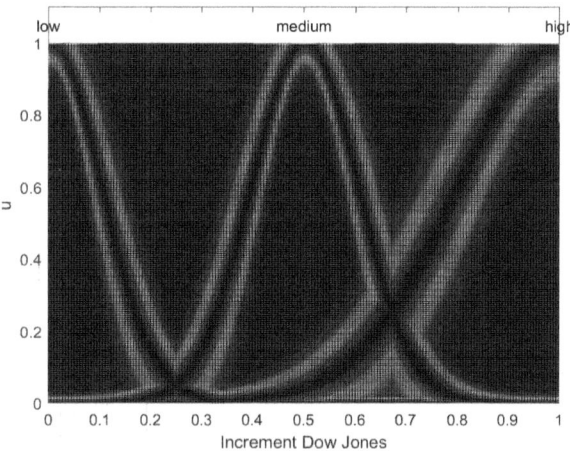

Fig. 5.5 MFs of COVID-19 variable

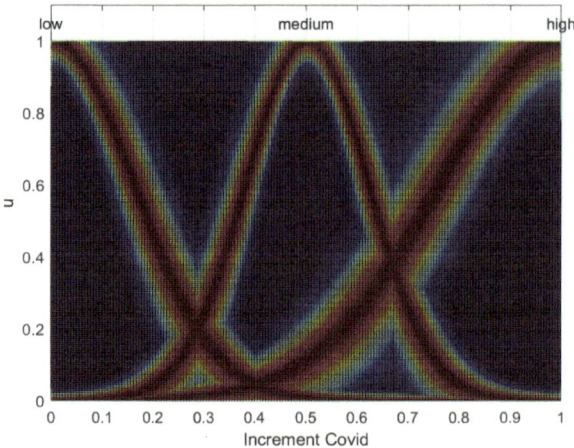

Fig. 5.6 MFs of output prediction

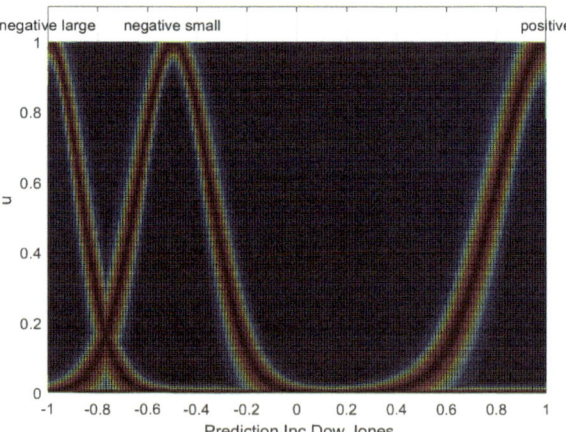

Fig. 5.7 A 3-D view of the model

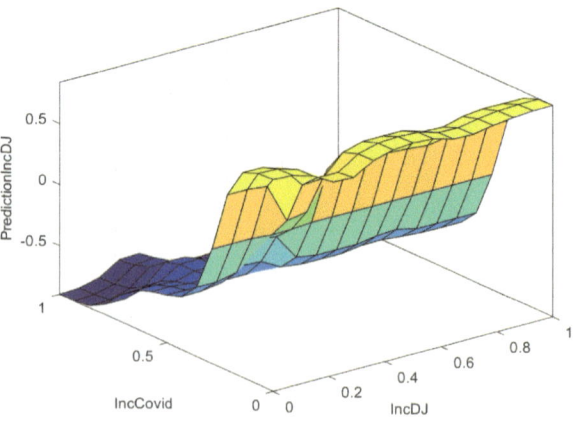

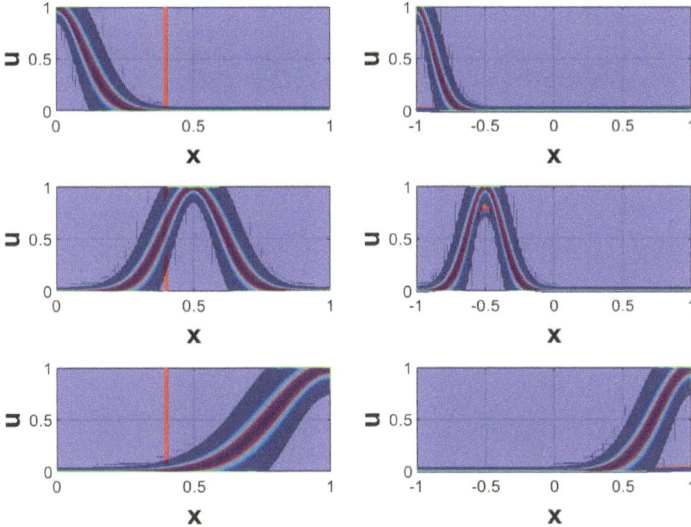

Fig. 5.8 Inference example

Fig. 5.9 Type reduction and defuzzification

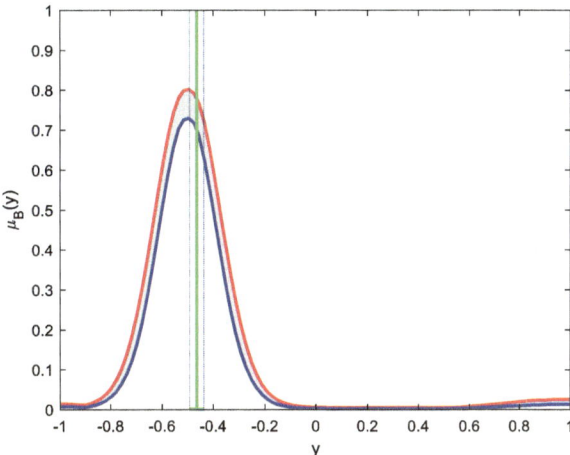

5.4 Results

The experiments were undertaken using a dataset from the Humanitarian Data Exchange (HDX) [15], which covers COVID-19 data of countries. Also, the Dow Jones (DJ) was collected (same time period) to be utilized [31]. We consider a period of 15 days that are utilized for testing (July 4 to July 18 of 2021). The idea is that the DJ forecast can be enhanced by considering COVID-19, as the Pandemic was supposed to be affecting the Economies of the countries.

The NN modules were trained with COVID-19 and DJ data from January of 2020 to May, 2021. Recurrent NNs were employed, with three delays, 300 epochs of backpropagation with momentum learning. There are 3 layers in the NNs.

Table 5.2 exhibits the forecasts of the NNs, and then the obtained increments of both series. Finally, the forecast calculated with the aggregator is presented (Dow Jones IT3) and the real value of DJ for comparison. Figure 5.10 depicts the prediction compared to the real DJ for the testing period from July 4 to 18, 2021.

Table 5.2 Forecasts of the NNs and fuzzy aggregation for July 4 to July 18 of 2021

DJ NN	COVID NN	DJ IT3	DJ Real
34,349.2571	33,849,760	34,948.4960	34,421.93
34,137.6330	33,861,363	35,536.1257	34,870.16
34,477.8324	33,895,756	35,113.1368	34,996.18
34,665.8612	33,921,173	34,697.2557	34,888.79
34,598.9918	33,946,079	35,257.7595	34,933.23
34,614.0874	34,015,922	34,894.7521	34,987.02
34,664.9170	34,015,183	34,247.8663	34,687.85
34,422.1593	34,023,764	33,628.0338	33,962.04
33,696.8697	34,068,368	33,999.7447	34,511.99
34,055.5272	34,101,607	34,436.4416	34,798.00
34,419.2452	34,141,916	34,820.1923	34,823.35
34,494.4152	34,182,819	34,439.2702	35,061.55
34,695.4936	34,301,941	34,862.2591	35,144.31
34,803.3353	34,297,819	35,488.4016	35,058.52
34,748.0242	34,317,020	34,927.8978	34,930.93

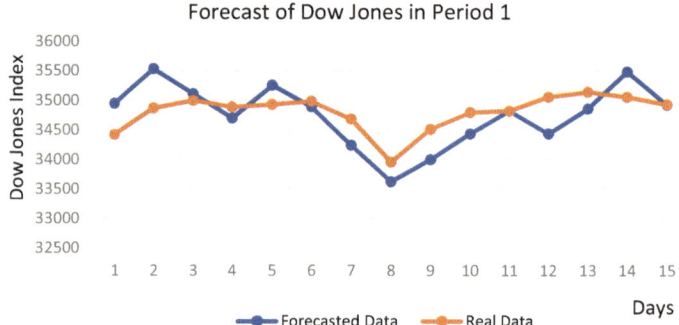

Fig. 5.10 Forecast of the DJ

5.5 Conclusions

Type-3 aggregation of NNs was outlined and the results surpass other methods. As future work we will consider other applications, as in [32–35]. We plan to optimize the type-3 systems for enhancing prediction accuracy. We can also consider other time series. Finally, we can optimize the structure of the NN models with the utilization of diverse metaheuristics, such as [36–39].

References

1. L.A. Zadeh, Knowledge representation in Fuzzy Logic. IEEE Trans. Knowl. Data Eng.Knowl. Data Eng. **1**, 89 (1989)
2. L.A. Zadeh, Fuzzy Logic. Computer **1**(4), 83–93 (1998)
3. J.M. Mendel, *Uncertain Rule-Based Fuzzy Logic Systems: Introduction and New Directions* (Prentice-Hall, Upper-Saddle River, NJ, 2001)
4. J. M. Mendel, *Uncertain Rule-Based Fuzzy Logic Systems: Introduction and New Directions*, 2nd edn. (Springer, 2017)
5. N.N. Karnik, J.M. Mendel, Operations on Type-2 fuzzy sets. Fuzzy Sets Syst. **122**, 327–348 (2001)
6. J.E. Moreno et al., Design of an interval Type-2 fuzzy model with justifiable uncertainty. Inf. Sci. **513**, 206–221 (2020)
7. J.M. Mendel, H. Hagras, W.-W. Tan, W.W. Melek, H. Ying, *Introduction to Type-2 Fuzzy Logic Control* (Wiley and IEEE Press, Hoboken, NJ, 2014)
8. F. Olivas, F. Valdez, O. Castillo, P. Melin, Dynamic parameter adaptation in particle swarm optimization using interval type-2 fuzzy logic. Soft. Comput.Comput. **20**(3), 1057–1070 (2016)
9. A. Sakalli, T. Kumbasar, J.M. Mendel, Towards systematic design of general type-2 fuzzy logic controllers: analysis, interpretation, and tuning. IEEE Trans. Fuzzy Syst. **29**(2), 226–239 (2021)
10. E. Ontiveros, P. Melin, O. Castillo, High order α-planes integration: a new approach to computational cost reduction of general type-2 fuzzy systems. Eng. Appl. Artif. Intell.Artif. Intell. **74**, 186–197 (2018)
11. O. Castillo, L. Amador-Angulo, A generalized type-2 fuzzy logic approach for dynamic parameter adaptation in bee colony optimization applied to fuzzy controller design. Inf. Sci. **460–461**, 476–496 (2018)
12. Y. Cao, A. Raise, A. Mohammadzadeh et al., Deep learned recurrent type-3 fuzzy system: application for renewable energy modeling/prediction. Energy Rep. (2021)
13. A. Mohammadzadeh, O. Castillo, S.S. Band et al., A novel fractional-order multiple-model type-3 fuzzy control for nonlinear systems with unmodeled dynamics. Int. J. Fuzzy Syst. (2021). https://doi.org/10.1007/s40815-021-01058-1
14. S.N. Qasem, A. Ahmadian, A. Mohammadzadeh, S. Rathinasamy, B. Pahlevanzadeh, A type-3 logic fuzzy system: optimized by a correntropy based Kalman filter with adaptive fuzzy kernel size Inform. Sci. **572**, 424–443 (2021)
15. The Humanitarian Data Exchange (HDX). https://data.humdata.org/dataset/novel-coronavirus-2019-ncov-cases. Accessed 31 Mar 2020
16. M.A. Shereen, S. Khan, A. Kazmi, N. Bashir, R. Siddique, COVID-19 infection: origin, transmission, and characteristics of human coronaviruses. J. Adv. Res. **24**, 91–98 (2020)
17. C. Sohrabi, Z. Alsafi, N. O'Neill, M. Khan, A. Kerwan, A. Al-Jabir, C. Iosifidis, R. Agha, World Health Organization declares global emergency: a review of the 2019 novel coronavirus (COVID-19). Int. J. Surg. **76**, 71–76 (2020)

18. I.D. Apostolopoulos, T. Bessiana, Covid-19: Automatic detection from X-Ray images utilizing Transfer Learning with Convolutional Neural Networks (2020). arXiv:2003.11617
19. S.A. Sarkodie, P.A. Owusu, Investigating the cases of novel coronavirus disease (COVID-19) in China using dynamic statistical techniques (2020). SSRN 3559456
20. B.R. Beck, B. Shin, Y. Choi, S. Park, K. Kang, Predicting commercially available antiviral drugs that may act on the novel coronavirus (SARS-CoV-2) through a drug-target interaction deep learning model. Comput. Struct. Biotechnol. J.. Struct. Biotechnol. J. **18**, 784–790 (2020)
21. L. Zhong, L. Mu, J. Li, J. Wang, Z. Yin, D. Liu, Early prediction of the 2019 novel coronavirus outbreak in the Mainland China based on simple mathematical model. IEEE Access **8**, 51761–51769 (2020)
22. M.N. Kamel Boulos, E.M. Geraghty, Geographical tracking and mapping of coronavirus disease COVID-19/severe acute respiratory syndrome coronavirus 2 (SARS-CoV-2) epidemic and associated events around the world: How 21st century GIS technologies are supporting the global fight against outbreaks and epidemics. Int. J. Health Geogr. **19**, 8 (2020). https://doi.org/10.1186/s12942-020-00202-8
23. P. Gao, H. Zhang, Z. Wu, J. Wang, Visualising the expansion and spread of coronavirus disease 2019 by cartograms. Environ Plan A (2020). https://doi.org/10.1177/0308518X20910162
24. A.S.R.S. Rao, J.A. Vazquez, Identification of COVID-19 can be quicker through artificial intelligence framework using a mobile phone-based survey in the populations when Cities/Towns are under quarantine. Infect. Control Hosp. Epidemiol.Epidemiol. (2020). https://doi.org/10.1017/ice.2020.61
25. P. Melin, J.C. Monica, D. Sanchez, O. Castillo, Analysis of spatial spread relationships of coronavirus (COVID-19) pandemic in the world using self organizing maps, Chaos, Solitons and Fractals, **138**, 109917 (2020). (pp. 1–7)
26. P. Melin, J.C. Monica, D. Sanchez, O. Castillo, Multiple ensemble neural network models with fuzzy response aggregation for predicting COVID-19 time series: the case of Mexico. Healthcare **8**, 181 (2020)
27. J.T. Rickard, J. Aisbett, G. Gibbon, Fuzzy subsethood for fuzzy sets of type-2 and generalized type-n. IEEE Trans. Fuzzy Syst. **17**(1), 50–60 (2009)
28. A. Mohammadzadeh, M.H. Sabzalian, W. Zhang, An interval type-3 fuzzy system and a new online fractional-order learning algorithm: theory and practice. IEEE Trans. Fuzzy Syst. **28**(9), 1940–1950 (2020)
29. Z. Liu, A. Mohammadzadeh, H. Turabieh, M. Mafarja, S.S. Band, A. Mosavi, A new online learned interval type-3 fuzzy control system for solar energy management systems. IEEE Access **9**, 10498–10508 (2021)
30. O. Castillo, J.R. Castro, P. Melin, *Interval Type-3 Fuzzy Systems: Theory and Design* (Springer, Cham, Switzerland, 2022)
31. Dow Jones time series data. https://m.mx.investing.com/indices/us-30-historical-data . Accessed 31 Mar 2022
32. L. Cervantes, O. Castillo, Type-2 fuzzy logic aggregation of multiple fuzzy controllers for airplane flight control. Inf. Sci. **324**, 247–256 (2015)
33. P. Melin, O. Castillo, An intelligent hybrid approach for industrial quality control combining neural networks, fuzzy logic and fractal theory. Inf. Sci. **177**, 1543–1557 (2007)
34. O. Castillo, J.R. Castro, P. Melin, A. Rodriguez-Diaz, Application of interval type-2 fuzzy neural networks in non-linear identification and time series prediction. Soft. Comput.Comput. **18**(6), 1213–1224 (2014)
35. E. Rubio, O. Castillo, F. Valdez, P. Melin, C.I. Gonzalez, G. Martinez, An extension of the fuzzy possibilistic clustering algorithm using type-2 fuzzy logic techniques. Adv. Fuzzy Syst. (2017). https://doi.org/10.1155/2017/7094046
36. O. Castillo, E. Lizzarraga, J. Soria, P. Melin, F. Valdez, New approach using ant colony optimization with ant set partition for fuzzy control design applied to the ball and beam system. Inf. Sci. **294**, 203–215 (2015)
37. L. Amador-Angulo, O. Mendoza, J.R. Castro, A. Rodriguez-Diaz, P. Melin, O. Castillo, Fuzzy sets in dynamic adaptation of parameters of a bee colony optimization for controlling the trajectory of an autonomous mobile robot. Sensors **16**(9), 1458 (2016)

38. F. Valdez, J.C. Vazquez, P. Melin, O. Castillo, Comparative study of the use of fuzzy logic in improving particle swarm optimization variants for mathematical functions using co-evolution. Appl. Soft Comput.Comput. **52**, 1070–1083 (2017)
39. B. González, F. Valdez, P. Melin, G. Prado-Arechiga, Fuzzy logic in the gravitational search algorithm for the optimization of modular neural networks in pattern recognition. Expert Syst. Appl. **42**(14), 5839–5847 (2015)

Chapter 6
Type-3 Fuzzy Aggregators for Neural Network Ensembles in Prediction

Abstract This chapter puts forward an approach for fuzzy aggregation in ensembles of NNs.

Keywords Type-3 fuzzy logic · Fuzzy aggregation

6.1 Introduction

The use of fuzzy logic enhances results in many areas [1, 2]. Type-1 evolved to type-2 since 2001 [3]. Later, these systems were applied in many areas [3–5]. Results show that interval type-2 surpasses type-1 when there are higher levels of noise [6–8]. Later, general type-2 was used to manage higher uncertainty levels, and remarkable results were achieved in some problems [9–11]. More recently, it has been accepted that type-3 can solve complicated problems. For this reason, we are proposing the terminology of type-3 by extending type-2 [12–14] and considering its applicability.

Recently, the spread of COVID-19 was noted. For the case of Europe several countries have been impacted with the spread of COVID-19 [15–20]. Countries of the American continent have also suffered from COVID-19 [21–23]. There are also recent works on predicting COVID-19 behavior in space and time [24, 25]. However, still prediction remains a challenging task [26–30], where different methods have been utilized. As a contrast to previous approaches, the contribution of this chapter is the proposal interval type-3 fuzzy systems as aggregators of NNs in prediction.

The chapter structure is: Sect. 6.2 is dedicated to the type-3 terminology, Sect. 6.3 describes the proposal, Sect. 6.4 highlights the results, and Sect. 6.5 offers conclusions.

© The Author(s), under exclusive license to Springer Nature Switzerland AG 2024   61
O. Castillo and P. Melin, *Type-3 Fuzzy Logic in Time Series Prediction*,
SpringerBriefs in Computational Intelligence,
https://doi.org/10.1007/978-3-031-59714-5_6

6.2   Type-3 Fuzzy Logic

We start by recalling the fuzzy set concept postulated by Zadeh [1], where the membership to a set is any number in the [0, 1] interval. A type-1 fuzzy set A is:

$$A = \{(x, \mu_A(x))| \text{ for all } x \text{ in } X\} \tag{6.1}$$

where x is an element of universe X, and $\mu_A(x)$ is a MF with numeric values in [0, 1]. Later, as an extension of type-1, the type-2 fuzzy set concept was proposed, allowing membership to be a type-1 fuzzy set [3, 4]. The goal is to achieve a better uncertainty representation. A type-2 fuzzy set $\tilde{A}$ is:

$$\tilde{A} = \left\{\left((x, u), \mu_{\tilde{A}}(x, u)\right)| \forall x \in X, \forall u \in J_x \subseteq [0, 1]\right\} \tag{6.2}$$

in which $0 \leq \mu_{\tilde{A}}(x, u) \leq 1$. In fact, $J_x \subseteq [0, 1]$ represents the primary membership domain of x, and $\mu_{\tilde{A}}(x, u)$ is a type-1 secondary set. Later, a type-3 fuzzy set was proposed as a generalization of type-2, by using primary, secondary and tertiary MFs for achieving a better management of uncertainty.

Definition 6.1 A type-3 fuzzy set (T3 FS) [31–33], denoted by $A^{(3)}$, uses a MF in the Cartesian product $X \times [0, 1] \times [0, 1]$ in [0, 1], where X is the universe of the primary variable of $A^{(3)}$, x. The MF of $\mu_{A^{(3)}}$ is formulated by $\mu_{A^{(3)}}(x, u, v)$ (or $\mu_{A^{(3)}}$) and it is called a T3 MF of the T3 FS,

$$\mu_{A^{(3)}} : X \times [0, 1] \times [0, 1] \rightarrow [0, 1]$$
$$A^{(3)} = \{(x, u(x), v(x, u), \mu_{A^{(3)}}(x, u, v))| x \in X, u \in U \subseteq [0, 1], v \in V \subseteq [0, 1]\} \tag{6.3}$$

where U is the universe for u and V is the universe for v. If the tertiary MF is uniformly equal to 1 then an IT3 FS is obtained.

Figure 6.1 depicts and IT3 FS with IT3MF $\tilde{\mu}(x, u)$, where $\underline{\mu}(x, u)$ is the LMF and $\overline{\mu}(x, u)$ is the UMF. The embedded secondary T1 MFs in x/ of $\underline{A}$ and $\overline{A}$ are $\underline{f}_{x\prime}(u)$ and $\overline{f}_{x\prime}(u)$.

We employed type-3 MFs that are scaled Gaussians. This MF is denoted as, $\tilde{\mu}_A(x, u) = $ ScaleGaussScaleGauss IT3MF, with Gaussian $FOU(\mathbb{A})$, with upper parameters $[\sigma, m]$ for the UMF and λ, ℓ for the LMF to form the $DOU = [\underline{\mu}(x), \overline{\mu}(x)]$. The vertical cuts $\mathbb{A}_{(x)}(u)$ form the $FOU(\mathbb{A})$, and are Gaussian IT2 MFs, $\mu_{\mathbb{A}_{(x)}}(u)$ with parameters $[\sigma_u, m(x)]$ for the UMF and LMF with λ, and ℓ. The IT3 MF, $\tilde{\mu}_{\mathbb{A}}(x, u) = $ ScaleGaussScaleGaussIT3MF $(x, \{\{[\sigma, m]\}, \lambda, \ell\})$ with expressions:

$$\overline{u}(x) = exp\left[-\frac{1}{2}\left(\frac{x - m}{\sigma}\right)^2\right] \tag{6.4}$$

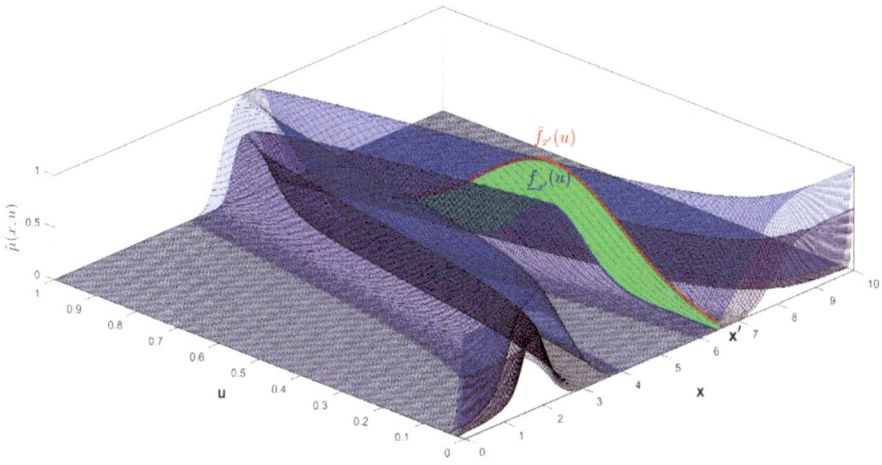

Fig. 6.1 Fuzzy set with $\tilde{\mu}(x, u)$

$$\underline{u}(x) = \lambda \cdot exp\left[-\frac{1}{2}\left(\frac{x - m}{\sigma^*}\right)^2\right] \tag{6.5}$$

where $\sigma^* = \sigma\sqrt{\frac{\ln(\ell)}{\ln(\varepsilon)}}$, ε is the machine epsilon. If $\ell = 0$, then $\sigma^* = \sigma$. Then $\overline{u}(x)$ and $\underline{u}(x)$ are the DOU upper and lower limits. The range, $\delta(u)$ and radius, σ_u are:

$$\delta(u) = \overline{u}(x) - \underline{u}(x) \tag{6.6}$$

$$\sigma_u = \frac{\delta(u)}{2\sqrt{3}} + \varepsilon \tag{6.7}$$

The apex or core, $m(x)$, of the IT3 MF $\tilde{\mu}(x, u)$, is postulated by:

$$m(x) = exp\left[-\frac{1}{2}\left(\frac{x - m}{\rho}\right)^2\right] \tag{6.8}$$

where $\rho = (\sigma + \sigma^*)/2$. Then, the vertical cuts with IT2 MF, $\mu_{\mathbb{A}(x)}(u) = [\underline{\mu}_{\mathbb{A}(x)}(u), \overline{\mu}_{\mathbb{A}(x)}(u)]$, are described by the equations:

$$\overline{\mu}_{\mathbb{A}(x)}(u) = exp\left[-\frac{1}{2}\left(\frac{u - u(x)}{\sigma_u}\right)^2\right] \tag{6.9}$$

$$\underline{\mu}_{\mathbb{A}(x)}(u) = \lambda \cdot exp\left[-\frac{1}{2}\left(\frac{u - u(x)}{\sigma_u^*}\right)^2\right] \tag{6.10}$$

where $\sigma_u^* = \sigma_u \sqrt{\frac{\ln(\ell)}{\ln(\varepsilon)}}$. If $\ell = 0$, then $\sigma_u^* = \sigma_u$.

6.3   Proposal

Figure 6.2 depicts the method in a diagram, where we note that the data enters the ensemble modules and predictions P_1 and P_2 are estimated with errors e_1 and e_2.

The fuzzy rules with two modules are:

1. If (e_1 is S) and (e_2 is S), then (w_1 is H)(w_2 is H).
2. If (e_1 is S) and (e_2 is M), then (w_1 is H)(w_2 is M).
3. If (e_1 is S) and (e_2 is H), then (w_1 is H)(w_2 is L).
4. If (e_1 is M) and (e_2 is S), then (w_1 is M)(w_2 is H).
5. If (e_1 is M) and (e_2 is M), then (w_1 is M)(w_2 is M).
6. If (e_1 is M) and (e_2 is H), then (w_1 is M)(w_2 is L).
7. If (e_1 is H) and (e_2 is S), then (w_1 is L)(w_2 is H).
8. If (e_1 is H) and (e_2 is M), then (w_1 is L)(w_2 is M).
9. If (e_1 is H) and (e_2 is H), then (w_1 is L)(w_2 is L).

Where small $= S$, medium $= M$, high $= H$ and $L = $ Low. The rule design is based on NN training knowledge. It is known that a high error implies the NN weight is low. Also, low error means the NN weight is high. The type-3 system (of Fig. 6.3) has as inputs the errors of each NN, e_1 and e_2, respectively. After type-reduction, the system estimates the weights (w_1 and w_2) for each NN to obtain a final prediction P that is expressed by:

$$P = \frac{w_1 P_1 + w_2 P_2}{w_1 + w_2} \tag{6.11}$$

In Table 6.1 we list the particular parameters of the MFs, which were found by experimentation. Basically, Table 6.1 lists the MF centers and standard deviations.

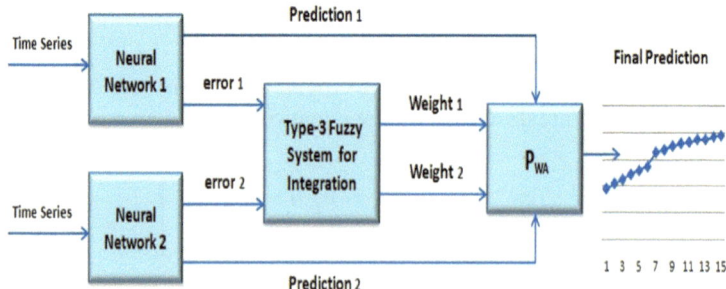

Fig. 6.2 Ensemble with type-3 fuzzy aggregation

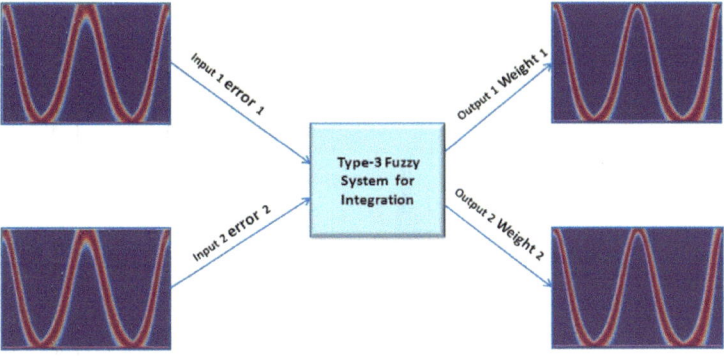

Fig. 6.3 Type-3 system to estimate the weights

The values of Table 6.1 are utilized in Eqs. (6.4)–(6.10) to specify the MFs in the rules.

Regarding the λ and ℓ parameters, they were found to be 0.8 and 0.2 for the inputs, and for the outputs of 0.9 and 0.6.

In Figs. 6.4 and 6.5 we depict the input MFs for both errors, respectively. In Figs. 6.6 and 6.7 we exhibit the output MFs for both weights. The MFs of these figures are obtained by plotting Eqs. (6.4)–(6.10) with values of Table 6.1. For example, for error e_1 (input 1) the parameters of the first three rows are employed in Eqs. (6.4)–(6.10) to generate Fig. 6.4.

In the outputs (Figs. 6.6 and 6.7), the weights are: low, medium and high. In Fig. 6.8 we show one view of the model surface. In Fig. 6.9 the view for w_2 with respect to errors is depicted.

Table 6.1 Parameter values for the MFs

Variable	MF	σ	m
Error 1	S	0.100	0.0
Error 1	M	0.120	0.5
Error 1	H	0.100	1.0
Error 2	S	0.100	0.0
Error 2	M	0.120	0.5
Error 2	H	0.100	1.0
Weight 1	S	0.100	0.0
Weight 1	M	0.110	0.5
Weight 1	H	0.100	1.0
Weight 2	S	0.100	0.0
Weight 2	M	0.110	0.5
Weight 2	H	0.100	1.0

Fig. 6.4 MFs of e_1

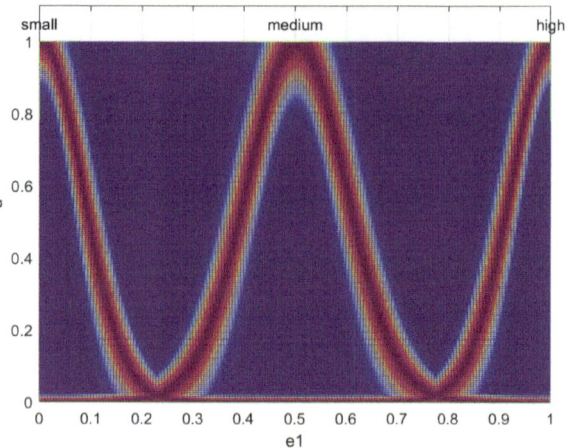

Fig. 6.5 MFs of e_2

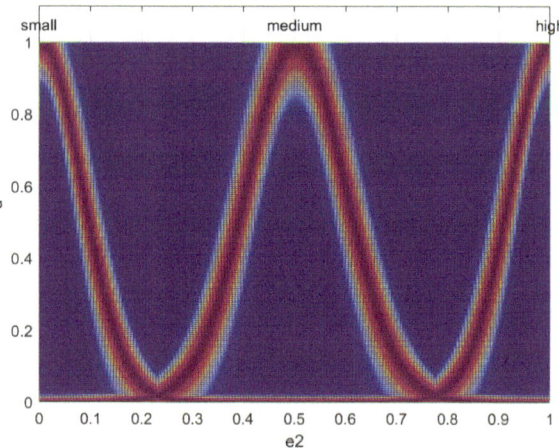

Fig. 6.6 MFs of w_1

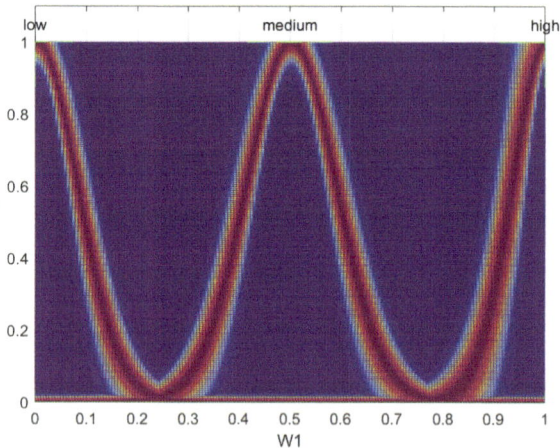

Fig. 6.7 MFs of w_2

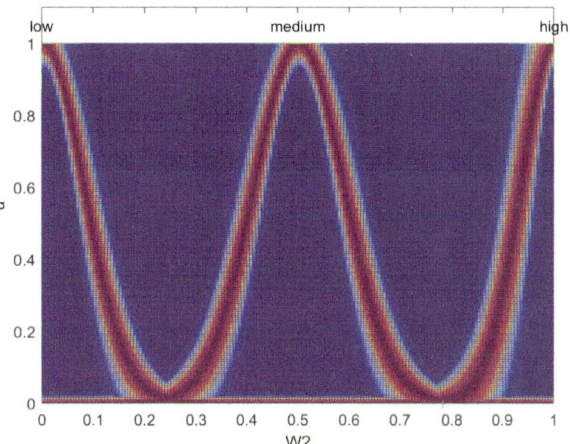

Fig. 6.8 View of the type-3 model for w_1

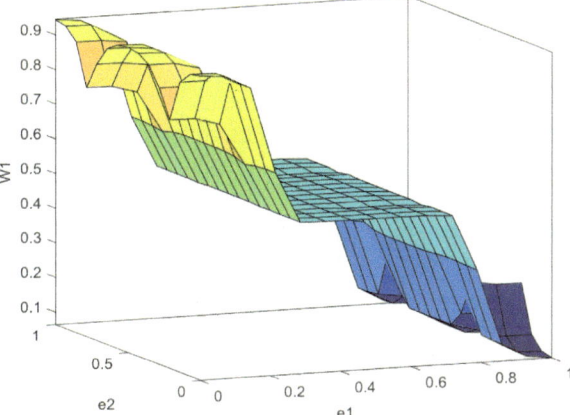

Fig. 6.9 Surface of the type-3 model for w_2

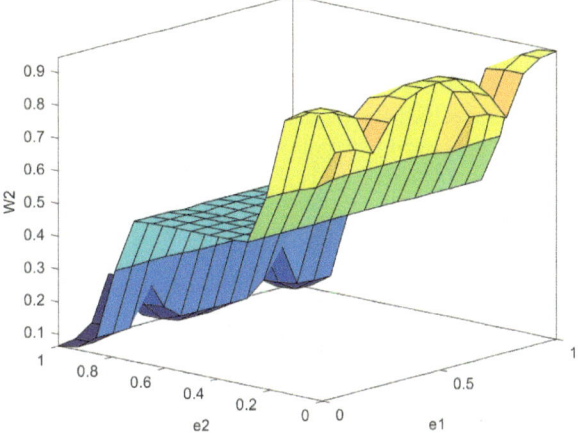

Table 6.2 Results for the ensemble modules

Country	e_1	e_2	w_1	w_2
F	1.00	0.82	0.165	0.186
G	0.956	0.11	0.094	0.913
J	1.00	0.86	0.113	0.118
P	0.095	1.00	0.920	0.087
U	0.70	1.00	0.496	0.139

6.4 Results

The experiments were performed with a dataset used from the Humanitarian Data Exchange (HDX) [15], which includes COVID-19 data from countries, the cases are from January 22, 2020 to January, 2022, where the last 15 days are for testing.

Table 6.2 shows the training errors of the ensemble modules (e_1 and e_2) and the corresponding weights. Results of Table 6.2 are for five countries. The NN modules were trained with COVID-19 data from January of 2020 to 2022, and the last 15 days are employed for testing. Recurrent NNs are employed, with 3 delays, 300 epochs of backpropagation with momentum learning. Three layers are employed in the NNs. In Table 6.3 results for France (F) are listed. In Table 6.4 we exhibit Germany (G) results and the prediction is depicted in Fig. 6.10. In Table 6.5 we list the results for Japan (J) and in Fig. 6.11 the prediction is depicted. Also, we depict in Table 6.6 and Fig. 6.12 the results for Poland (P). Lastly, Table 6.7 lists the USA (U) predictions.

According to Table 6.3 the results for France prediction are relatively close to the real values.

Germany prediction results are good according to Table 6.4 and is visualized in Fig. 6.10, where both forecasted and real values are plotted.

The Japan prediction results are very good according to Table 6.5 and evident in Fig. 6.11.

According to Table 6.6, Poland predictions for 15 days are very close to real values and this is evident in Fig. 6.12.

Lastly, in Table 6.7 we notice that USA predicted and real values are close.

In Table 6.8 we summarize a prediction comparison for 12 countries in which type-3 is better in 11 out of 12 cases.

6.5 Conclusions

In this chapter, fuzzy aggregation in ensembles of NNs has been described. As future work we plan to deal with different applications, as in [35–40]. Also, we can consider, general type-3 models instead of interval type-3, as envisioned in [41–44]. Finally, we can optimize the type-3 system for enhancing the results, such as the ones in [45–51].

Table 6.3 Prediction and comparison with real values for France

P_1	P_2	P_{WA}	P_{Real}
6,605,410.19	7,008,247.76	6,819,020.56	7,075,244
6,614,139.24	7,017,080.51	6,827,804.6	7,079,005
6,622,399.29	7,019,209.37	6,832,813.5	7,093,651
6,627,109.31	7,030,190.08	6,840,848.64	7,106,147
6,638,009.1	7,038,745.08	6,850,505.07	7,109,125
6,646,165.48	7,040,257.54	6,855,138.42	7,128,903
6,651,161.41	7,055,100.1	6,865,355.67	7,149,118
6,666,269.19	7,069,013.04	6,879,829.87	7,168,026
6,681,418.33	7,081,862.44	6,893,759.54	7,188,721
6,696,029.19	7,096,088.87	6,908,166.55	7,211,399
6,711,976.63	7,111,599.88	6,923,882.57	7,231,148
6,728,648.35	7,124,729.52	6,938,676.05	7,235,966
6,741,361.56	7,126,988.88	6,945,845.95	7,266,361
6,748,887.39	7,149,216.18	6,961,167.45	7,296,757
6,771,633.28	7,169,428.36	6,982,569.8	7,330,086

Table 6.4 Prediction and comparison with real values for Germany

P_1	P_2	P_{WA}	P_{Real}
4,874,657.83	8,186,604.82	7,874,981.02	7,866,784
4,877,993.44	8,284,832.8	7,964,280.49	7,943,959
4,880,960.91	8,372,079.56	8,043,597.35	7,988,210
4,882,089.4	8,414,449.58	8,082,086.92	8,021,339
4,882,788.23	8,444,313.88	8,109,207.02	8,104,157
4,886,304.01	8,546,418.12	8,202,034.98	8,222,262
4,891,326.88	8,696,895.89	8,338,826.79	8,361,262
4,896,263.17	8,864,841.73	8,491,434.92	8,502,132
4,900,498.6	9,027,704.27	8,639,372.11	8,635,461
4,903,941.72	9,175,951.89	8,773,994.95	8,716,804
4,905,466.24	9,255,829.58	8,846,500.33	8,773,030
4,906,195.81	9,304,413.44	8,890,581.54	8,909,503
4,909,695.04	9,465,430.18	9,036,777.33	9,088,672
4,914,154.61	9,684,141.65	9,235,329.67	9,317,280
4,918,690.59	9,948,751.83	9,475,469.25	9,477,603

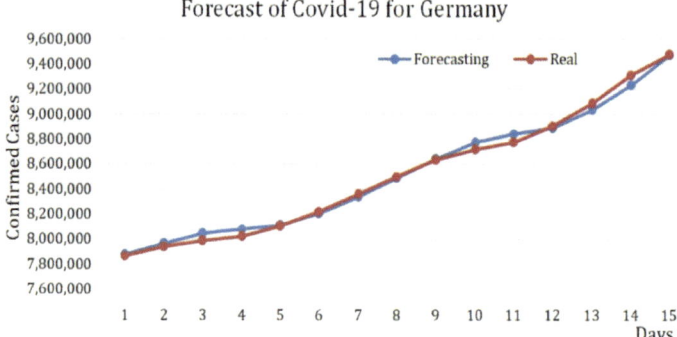

Fig. 6.10 Comparison of Germany prediction with real values

Table 6.5 Prediction and comparison with real values for Japan

P_1	P_2	P_{WA}	P_{Real}
1,421,031.66	1,168,372.61	1,292,086.17	1,209,082
1,435,781.23	1,191,360.06	1,311,039.97	1,234,423
1,449,848.17	1,213,401.89	1,329,176.93	1,260,415
1,462,806.61	1,235,123.08	1,346,607.47	1,286,077
1,474,669.31	1,255,662.63	1,362,898.43	1,308,422
1,484,070.3	1,271,603.87	1,375,637.26	1,325,340
1,517,540.24	1,340,270.56	1,427,069.99	1,420,783
1,525,625.61	1,357,837.38	1,439,994.25	1,443,642
1,532,582.42	1,373,612.4	1,451,451.47	1,462,979
1,537,962.41	1,385,600.95	1,460,204.16	1,476,653
1,541,193.63	1,392,303.61	1,465,207.05	1,494,372
1,545,591.19	1,406,220.89	1,474,463.04	1,514,400
1,545,591.19	1,406,220.89	1,474,463.04	1,514,400
1,551,355.7	1,421,248.78	1,484,955.15	1,532,616
1,556,035.67	1,432,831.96	1,493,158.2	1,549,337

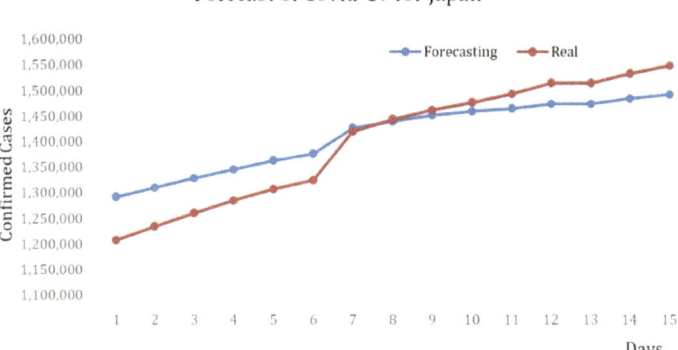

Fig. 6.11 Prediction comparison with respect to Japan real values

Table 6.6 Prediction and comparison with real values for Poland

P_1	P_2	P_{WA}	P_{Real}
3,995,819.87	3,621,223.9	3,963,176.88	3,881,349
4,003,077.22	3,637,991.57	3,971,262.98	3,903,445
4,008,232.79	3,647,846.54	3,976,828.06	3,923,472
4,013,534.37	3,659,477.29	3,982,681.17	3,942,864
4,018,723.82	3,671,368.56	3,988,454.63	3,958,840
4,022,446.88	3,679,114.97	3,992,528.3	3,968,450
4,024,370.78	3,682,395.61	3,994,570.43	3,982,257
4,028,881.2	3,694,834.14	3,999,771.72	4,000,270
4,033,453.12	3,706,100.13	4,004,926.97	4,017,420
4,037,113.47	3,714,046.58	4,008,960.82	4,032,796
4,040,580.05	3,722,064.06	4,012,823.97	4,043,585
4,042,683.73	3,726,273.81	4,015,111.18	4,049,838
4,043,955.25	3,728,860.68	4,016,497.32	4,054,865
4,045,314.96	3,732,474.08	4,018,053.42	4,064,715
4,048,172.85	3,740,829.22	4,021,390.35	4,080,282

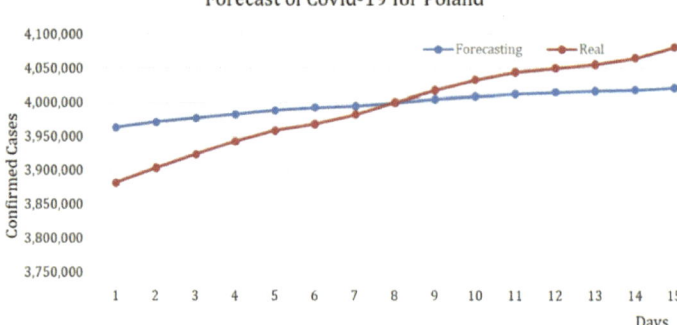

Fig. 6.12 Comparison of Poland prediction

Table 6.7 Prediction and comparison for USA

P1	P_2	P_{WA}	P_{Real}
61,326,847	60,718,991.9	61,193,732.4	64,210,668
62,011,425.7	61,319,440.6	61,859,887.4	65,069,619
62,675,140.7	61,930,734.2	62,512,122.6	65,463,200
62,750,945.4	62,101,842.9	62,608,798	65,930,556
62,982,269.4	62,412,455.5	62,857,485.4	66,590,148
63,622,278.3	62,921,873.6	63,468,896.2	67,693,339
64,773,747.8	63,794,454.8	64,559,291.7	68,684,431
65,636,754.3	64,470,771.5	65,381,414.9	69,388,781
65,982,382.4	64,869,656.6	65,738,705.8	70,206,083
66,548,711.7	65,413,568.4	66,300,125.8	70,490,987
66,527,725.9	65,492,563	66,301,034.9	70,845,794
66,625,713.4	65,702,223.5	66,423,477.7	71,741,698
67,597,051.5	66,400,650.5	67,335,050.8	72,257,016
67,982,681.7	66,685,204.4	67,698,546.2	72,910,136
68,386,962.5	67,101,527.1	68,105,464.1	73,427,335

Table 6.8 Prediction comparison for 12 cases with Mean Squared Error (MSE)

Country	Comparison					
	Type-2 [34]			Type-3 (This chapter)		
	Best	Avg	Worst	Best	Avg	Worst
Brazil	3.04×10^{-6}	**0.0197**	0.1420	1.84×10^{-6}	0.02460	0.1060
China	1.84×10^{-3}	0.0698	0.2960	5.22×10^{-4}	**0.02970**	0.1610
France	6.07×10^{-6}	0.0206	0.1940	4.13×10^{-6}	**0.00711**	0.0606
Germany	8.02×10^{-4}	0.0855	0.4110	4.09×10^{-5}	**0.03020**	0.1010
India	1.56×10^{-7}	0.0089	0.1540	3.05×10^{-7}	**0.00301**	0.0205
Iran	1.12×10^{-5}	0.0178	0.0982	7.56×10^{-7}	**0.01430**	0.1050
Italy	7.57×10^{-6}	0.0454	0.2920	1.24×10^{-5}	**0.01760**	0.0832
Mexico	2.86×10^{-5}	0.0091	0.1810	1.48×10^{-5}	**0.00149**	0.0300
Poland	8.31×10^{-5}	0.0205	0.4280	5.71×10^{-5}	**0.00608**	0.0585
Spain	5.82×10^{-4}	0.0009	0.0015	5.56×10^{-4}	**0.00083**	0.00154
United Kingdom	2.80×10^{-5}	0.0070	0.1620	2.69×10^{-4}	**0.01260**	0.0998
USA	3.15×10^{-6}	0.0080	0.0604	6.26×10^{-7}	**0.00532**	0.0949

References

1. L.A. Zadeh, Knowledge representation in Fuzzy Logic. IEEE Trans. Knowl. Data Eng. **1**, 89 (1989)
2. L.A. Zadeh, Fuzzy Logic. Computer **1**(4), 83–93 (1998)
3. J.M. Mendel, *Uncertain Rule-Based Fuzzy Logic Systems: Introduction and New Directions* (Prentice-Hall, Upper-Saddle River, NJ, 2001)
4. J.M. Mendel, *Uncertain Rule-Based Fuzzy Logic Systems: Introduction and New Directions*, 2nd edn. (Springer, 2017)
5. N.N. Karnik, J.M. Mendel, Operations on type-2 fuzzy sets. Fuzzy Sets Syst. **122**, 327–348 (2001)
6. J.E. Moreno et al., Design of an interval type-2 fuzzy model with justifiable uncertainty. Inf. Sci. **513**, 206–221 (2020)
7. J.M. Mendel, H. Hagras, W.-W. Tan, W.W. Melek, H. Ying, *Introduction to Type-2 Fuzzy Logic Control* (Wiley and IEEE Press, Hoboken, NJ, 2014)
8. F. Olivas, F. Valdez, O. Castillo, P. Melin, Dynamic parameter adaptation in particle swarm optimization using interval type-2 fuzzy logic. Soft. Comput. **20**(3), 1057–1070 (2016)
9. A. Sakalli, T. Kumbasar, J.M. Mendel, Towards systematic design of general type-2 fuzzy logic controllers: analysis, interpretation, and tuning. IEEE Trans. Fuzzy Syst. **29**(2), 226–239 (2021)
10. E. Ontiveros, P. Melin, O. Castillo, High order α-planes integration: a new approach to computational cost reduction of general type-2 fuzzy systems. Eng. Appl. Artif. Intell. **74**, 186–197 (2018)
11. O. Castillo, L. Amador-Angulo, A generalized type-2 fuzzy logic approach for dynamic parameter adaptation in bee colony optimization applied to fuzzy controller design. Inf. Sci. **460–461**, 476–496 (2018)
12. Y. Cao, A. Raise, A. Mohammadzadeh, et al., Deep learned recurrent type-3 fuzzy system: application for renewable energy modeling / prediction. Energy Rep. (2021)

13. A. Mohammadzadeh, O. Castillo, S.S. Band et al., A novel fractional-order multiple-model type-3 fuzzy control for nonlinear systems with unmodeled dynamics. Int. J. Fuzzy Syst. (2021). https://doi.org/10.1007/s40815-021-01058-1

14. S.N. Qasem, A. Ahmadian, A. Mohammadzadeh, S. Rathinasamy, B. Pahlevanzadeh, A type-3 logic fuzzy system: optimized by a correntropy based Kalman filter with adaptive fuzzy kernel size. Inf. Sci. **572**, 424–443 (2021)

15. The Humanitarian Data Exchange (HDX). https://data.humdata.org/dataset/novel-coronavirus-2019-ncov-cases. Accessed 31 Mar 2022

16. M.A. Shereen, S. Khan, A. Kazmi, N. Bashir, R. Siddique, COVID-19 infection: origin, transmission, and characteristics of human coronaviruses. J. Adv. Res. **24**, 91–98 (2020)

17. C. Sohrabi, Z. Alsafi, N. O'Neill, M. Khan, A. Kerwan, A. Al-Jabir, C. Iosifidis, R. Agha, World Health Organization declares global emergency: a review of the 2019 Novel coronavirus (COVID-19). Int. J. Surg. **76**, 71–76 (2020)

18. I.D. Apostolopoulos, T. Bessiana, Covid-19: automatic detection from X-Ray images utilizing transfer learning with convolutional neural networks (2020). arXiv:2003.11617

19. S.A. Sarkodie, P.A. Owusu, Investigating the cases of novel coronavirus disease (COVID-19) in China using dynamic statistical techniques (2020). SSRN 3559456

20. B.R. Beck, B. Shin, Y. Choi, S. Park, K. Kang, Predicting commercially available antiviral drugs that may act on the novel coronavirus (SARS-CoV-2) through a drug-target interaction deep learning model. Comput. Struct. Biotechnol. J. **18**, 784–790 (2020)

21. L. Zhong, L. Mu, J. Li, J. Wang, Z. Yin, D. Liu, Early prediction of the 2019 novel coronavirus outbreak in the Mainland China based on simple mathematical model. IEEE Access **8**, 51761–51769 (2020)

22. M.N. Kamel Boulos, E.M. Geraghty, Geographical tracking and mapping of coronavirus disease COVID-19/severe acute respiratory syndrome coronavirus 2 (SARS-CoV-2) epidemic and associated events around the world: How 21st century GIS technologies are supporting the global fight against outbreaks and epidemics. Int J Health Geogr. **19**, 8 (2020). https://doi.org/10.1186/s12942-020-00202-8

23. P. Gao, H. Zhang, Z. Wu, J. Wang, Visualising the expansion and spread of coronavirus disease 2019 by cartograms. Environ. Plan. A (2020). https://doi.org/10.1177/0308518X20910162

24. A.S.R.S. Rao, J.A. Vazquez, Identification of COVID-19 can be quicker through artificial intelligence framework using a mobile phone-based survey in the populations when Cities/Towns are under quarantine. Infect. Control Hospital Epidemiol. (2020). https://doi.org/10.1017/ice.2020.61

25. P. Melin, J.C. Monica, D. Sanchez, O. Castillo, Analysis of spatial spread relationships of coronavirus (COVID-19) pandemic in the world using self organizing maps. Chaos Solitons Fractals **138**(109917), 1–7 (2020)

26. P. Melin, J.C. Monica, D. Sanchez, O. Castillo, Multiple ensemble neural network models with fuzzy response aggregation for predicting COVID-19 time series: the case of Mexico. Healthcare **8**, 181 (2020)

27. Z. Jin, J.Y. Liu, R. Feng, L. Ji, Z.L. Jin, H.B. Li, Drug treatment of coronavirus disease 2019 (COVID-19) in China. Eur. J. Pharmacol. **883**, 1–7 (2020)

28. S. Khalilpourazari, H.H. Doulabi, A.Ö. Çiftçioglu, G.W. Weber, Gradient-based grey wolf optimizer with Gaussian walk: application in modelling and prediction of the COVID-19 pandemic. Expert Syst. Appl. **177**, 1–23 (2021)

29. Y. Kuvvetli, M. Deveci, T. Paksoy, H. Garg, A predictive analytics model for COVID-19 pandemic using artificial neural networks. Decis. Anal. J. **1**, 1–13 (2021)

30. D. Liu, W. Ding, Z.S. Dong, W. Pedrycz, Optimizing deep neural networks to predict the effect of social distancing on COVID-19 spread. Comput. Ind. Eng. **166**, 1–17 (2022)

31. J.T. Rickard, J. Aisbett, G. Gibbon, Fuzzy subsethood for fuzzy sets of type-2 and generalized type-n. IEEE Trans. Fuzzy Syst. **17**(1), 50–60 (2009)

32. A. Mohammadzadeh, M.H. Sabzalian, W. Zhang, An interval type-3 fuzzy system and a new online fractional-order learning algorithm: theory and practice. IEEE Trans. Fuzzy Syst. **28**(9), 1940–1950 (2020)

33. Z. Liu, A. Mohammadzadeh, H. Turabieh, M. Mafarja, S.S. Band, A. Mosavi, A new online learned interval type-3 fuzzy control system for solar energy management systems. IEEE Access **9**, 10498–10508 (2021)
34. P. Melin, D. Sánchez, J.C. Monica, O. Castillo, Optimization using the firefly algorithm of ensemble neural networks with type-2 fuzzy integration for COVID-19 time series prediction. Soft. Comput. **1**, 1–38 (2021)
35. L. Cervantes, O. Castillo, Type-2 fuzzy logic aggregation of multiple fuzzy controllers for airplane flight control. Inf. Sci. **324**, 247–256 (2015)
36. P. Melin, O. Castillo, An intelligent hybrid approach for industrial quality control combining neural networks, fuzzy logic and fractal theory. Inf. Sci. **177**, 1543–1557 (2007)
37. O. Castillo, J.R. Castro, P. Melin, A. Rodriguez-Diaz, Application of interval type-2 fuzzy neural networks in non-linear identification and time series prediction. Soft. Comput. **18**(6), 1213–1224 (2014)
38. E. Rubio, O. Castillo, F. Valdez, P. Melin, C.I. Gonzalez, G. Martinez, An extension of the fuzzy possibilistic clustering algorithm using type-2 fuzzy logic techniques. Adv. Fuzzy Syst. (2017). https://doi.org/10.1155/2017/7094046
39. M.W. Tian, A. Mohammadzadeh, J. Tavoosi, S. Mobayen, J.H. Asad, O. Castillo, A.R. Várkonyi-Kóczy, A deep-learned type-3 fuzzy system and its application in modeling problems. Acta Polytech. Hung. **19**(2) (2022)
40. A.A. Aly, B.F. Felemban, A. Mohammadzadeh, O. Castillo, A. Bartoszewicz, Frequency regulation system: a deep learning identification, type-3 fuzzy control and LMI stability analysis. Energies **14**(22), 7801 (2021)
41. O. Castillo, J.R. Castro, P. Melin, *Interval Type-3 Fuzzy Systems: Theory and Design.* (Springer, 2022)
42. O. Castillo, P. Melin, Review of type-3 fuzzy control, in *Type-3 Fuzzy Logic in Intelligent Control.* SpringerBriefs in Applied Sciences and Technology. (Springer, Cham, 2023). https://doi.org/10.1007/978-3-031-46088-3_3
43. O. Castillo, P. Melin, Type-3 fuzzy theory, in *Type-3 Fuzzy Logic in Intelligent Control.* SpringerBriefs in Applied Sciences and Technology. (Springer, Cham, 2023). https://doi.org/10.1007/978-3-031-46088-3_2
44. O Castillo, P. Melin, Approach for type-3 fuzzy control, in *Type-3 Fuzzy Logic in Intelligent Control.* SpringerBriefs in Applied Sciences and Technology. (Springer, Cham, 2023). https://doi.org/10.1007/978-3-031-46088-3_4
45. O. Castillo, E. Lizzarraga, J. Soria, P. Melin, F. Valdez, New approach using ant colony optimization with ant set partition for fuzzy control design applied to the ball and beam system. Inf. Sci. **294**, 203–215 (2015)
46. L. Amador-Angulo, O. Mendoza, J.R. Castro, A. Rodriguez-Diaz, P. Melin, O. Castillo, Fuzzy sets in dynamic adaptation of parameters of a bee colony optimization for controlling the trajectory of an autonomous mobile robot. Sensors **16**(9), 1458 (2016)
47. F. Valdez, J.C. Vazquez, P. Melin, O. Castillo, Comparative study of the use of fuzzy logic in improving particle swarm optimization variants for mathematical functions using co-evolution. Appl. Soft Comput. **52**, 1070–1083 (2017)
48. B. González, F. Valdez, P. Melin, G. Prado-Arechiga, Fuzzy logic in the gravitational search algorithm for the optimization of modular neural networks in pattern recognition. Expert Syst. Appl. **42**(14), 5839–5847 (2015)
49. F. Valdez, H. Carreon-Ortiz, O. Castillo, CMOA—Continuous Mycorrhiza Optimization Algorithm, in *Mycorrhiza Optimization Algorithm.* SpringerBriefs in Applied Sciences and Technology. (Springer, Cham, 2023). https://doi.org/10.1007/978-3-031-47369-2_5
50. F. Valdez, H. Carreon-Ortiz, O. Castillo, DMOA—Discrete Mycorrhiza Optimization Algorithm, in *Mycorrhiza Optimization Algorithm.* SpringerBriefs in Applied Sciences and Technology. (Springer, Cham, 2023). https://doi.org/10.1007/978-3-031-47369-2_6
51. M.H.F. Zarandi, A.A.S. Asl, S. Sotudian, O. Castillo, A state of the art review of intelligent scheduling. Artif. Intell. Rev. **53**, 501–593 (2020)

Chapter 7
Optimal Type-3 Fuzzy Systems and Ensembles of Neural Networks Using the Firefly Algorithm

Abstract In this chapter, the COVID-19 information is employed to perform times series prediction. We put forward the optimal design of ensemble neural networks (ENNs).

Keywords Prediction · Ensembles · Firefly algorithm · Type-3 fuzzy logic

7.1 Introduction

The pandemic has affected population worldwide. Studies regarding this disease and its sequels have continued emerging [1–3]. For example, in [4], a study of patients with uncommon illnesses from Northern India is presented, where authors demonstrated that if patients have illnesses, such as tuberculosis, they can havesevere COVID-19 because in India diseases, such as tuberculosis, have not been eradicated. Female patients from Osaka, Japan, diagnosed with COVID-19 are studied in [5] to evaluate pregnant women and determine if they are likely to have severe COVID-19. The authors concluded that they are more likely to be hospitalized but will not have severe COVID-19. In [6], the use of dexamethasone is studied, as it reduces the mortality of COVID-19 patients who need oxygen or mechanical ventilation. The authors conclude that prolonged use of this drug increases the risk of developing a respiratory tract infection. Intelligent techniques have also a relevant role in the study of this disease. Deep neural networks (DNNs) and an improved particle swarm optimization are proposed to predict the social distancing effect on COVID-19 in [7]. In [8], a predictive model using ANNs to predict ten days of confirmed and death COVID-19 cases is proposed. A hybrid method utilizing intelligence techniques, such as recurrent and convolutional NNs, to predict confirmed cases of the seven most affected states in India is presented in [9]. In [10], an enhancement of the Grey Wolf Optimizer to accelerate the convergence to predict USA COVID-19 cases is proposed. ENN as an improvement of a conventional NN, because it can predict future information based on a data learned by the ENN modules [11–14]. In [15], ENNs were implemented to predict COVID-19 cases, and the architectures

© The Author(s), under exclusive license to Springer Nature Switzerland AG 2024   77
O. Castillo and P. Melin, *Type-3 Fuzzy Logic in Time Series Prediction*,
SpringerBriefs in Computational Intelligence,
https://doi.org/10.1007/978-3-031-59714-5_7

were designed using FA. The simulations exhibited better results using Type-2 FWA. Type-3 has been applied to complex problems where it is combined with other techniques [16, 17]. In [18], a deep learned model with type-3 to forecast renewable energies is proposed. The contribution of this chapter resides on the type-3 mixture of responses of the ENNs. Additionally, the Type-3 FISs and NNs are optimally designed using a firefly algorithm (FA).

The chapter is structured as: A description of the used techniques is outlined in Sect. 7.2. The explanation of the method is contained in Sect. 7.3. Experiments are detailed in Sect. 7.4. Conclusions are highlighted in Sect. 7.5.

7.2   Intelligent Techniques

The intelligent techniques applied in the method are presented.

7.2.1   NNs

Humans have abilities that differ from other species. Among these, we can find their abilities to learn and recognize. The brain accomplishes the tasks by recognizing and learning. The next step consists of recognizing shapes, like letters or numbers [19, 20]. The previously discussed abilities are simulated by an artificial intelligence technique called artificial neural networks (ANNs). ANNs can simulate complex problems by adjusting its parameters (weights), which in the learning process, store knowledge [21–23]. In Fig. 7.1, a representation of an ANN is shown [24, 25].

The enhancement of an ANN is to create an ensemble NN. The ensembles are composed of several ANNs that learn the same information, forming different experts for the same task [13]. A global result is enabled by combining the NN responses using an aggregation technique. In this work, this kind of ANN is applied to obtain an individual prediction to combine them with an aggregator to achieve a final prediction [12, 14]. In Fig. 7.2, an ENN with 3 modules is depicted.

Fig. 7.1 Artificial neural network

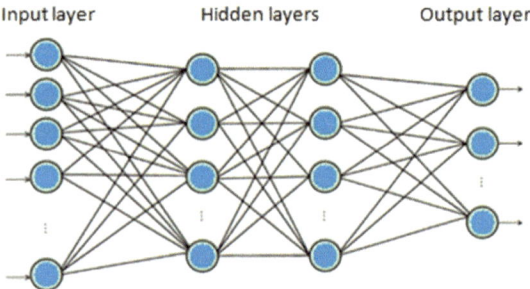

Input layer   Hidden layers   Output layer

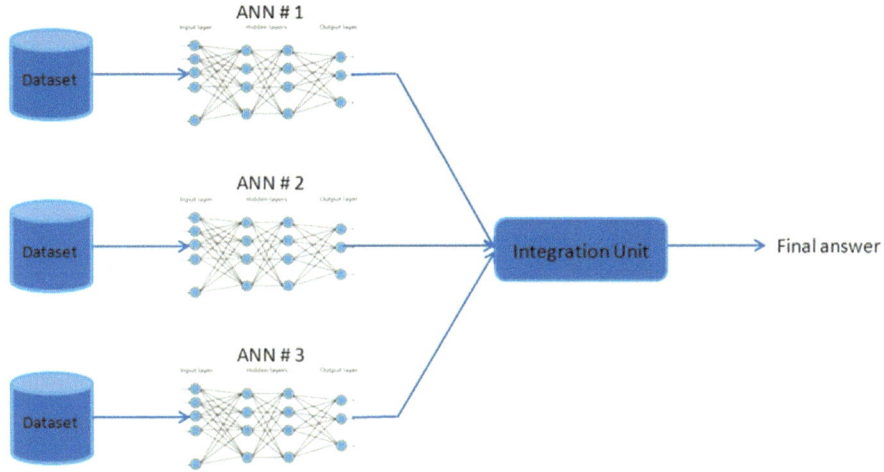

Fig. 7.2 Ensemble NN

7.2.2 *Type-3 Fuzzy Logic*

In [26], Zadeh proposed a useful intelligent technique: Type-1 fuzzy logic (FL). Type-2 FL was later proposed in [27]. In Type-2 FL [28–30], the MF is defined by a fuzzy set in [0,1]. A type-2 FS can be postulated as:

$$\tilde{A} = \left\{ \left((x, u), \mu_{\tilde{A}}(x, u) \right) | \forall_x \in X, \forall_u \in J_x \subseteq [0, 1], \mu_{\tilde{A}}(x, u) \in [0, 1] \right\} \quad (7.1)$$

where X represents the domain of x. In this case, there exist a primary MF and a secondary MF. If $\mu_{\tilde{A}}(x, u) = 1$, $\forall_u \in J_x \subseteq [0, 1]$, there is an IT2 MF as in Fig. 7.3, where there is a uniform shading FOU with its upper $\overline{\mu}_{\tilde{A}}(x)$ and lower $\underline{\mu}_{\tilde{A}}(x)$ MF [31]. An interval Type-2 FS is defined as:

$$\tilde{A} = \{((x, u), 1) | \forall x \in X, \forall_u \in J_x \subseteq [0, 1]\} \quad (7.2)$$

A Type-3 fuzzy set (T3 FS) [32, 33] denoted as $A^{(3)}$, is represented by a trivariate function, called MF of $A^{(3)}$,, in the cartesian product (Eq. 7.3), where X is the universe for the primary variable of $A^{(3)}$, x. The MF of $\mu_{A^{(3)}}$ is denoted by $\mu_{A^{(3)}}(x, u, v)$, and is a Type-3 MF:

$$\mu_{A^{(3)}} : X \times [0, 1] \times [0, 1] \rightarrow [0, 1] \quad (7.3)$$

$$A^{(3)} = \{(x, u(x), v(x, u), \mu_{A^{(3)}}(x, u, v)) | x \in X, u \in U \subseteq [0, 1], v \in V \subseteq [0, 1]\} \quad (7.4)$$

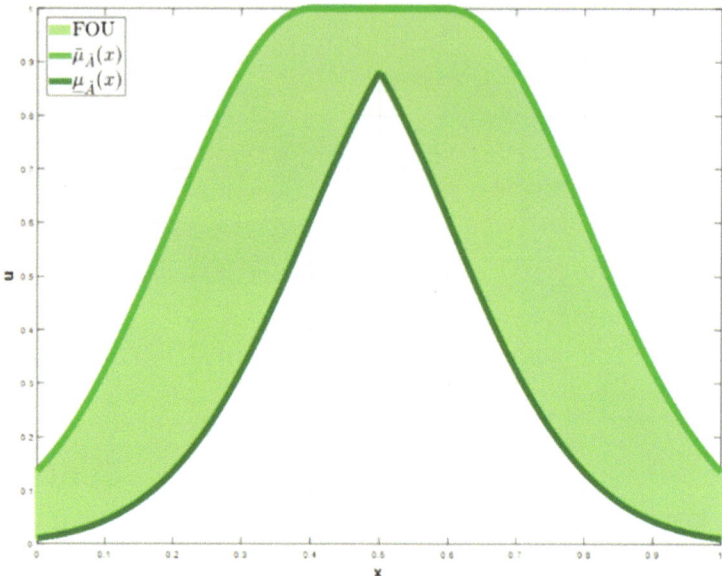

Fig. 7.3 Gaussian type-2 MF

where for the secondary variable u its universe is U, and V for the tertiary variable v. A Gaussian MF $\tilde{\mu}_{\mathbb{A}}(x, u) = $ ScaleGaussScaleGaussIT3MF with Gaussian $FOU(\mathbb{A})$ has for the (UMF) as parameters $[\sigma, m]$, and for the LMF λ and ℓ to form $DOU = [\underline{\mu}(x), \overline{\mu}(x)]$. This membership function is represented as:

$$\tilde{\mu}_{\mathbb{A}}(x, u) = \textbf{ScaleGaussScaleGaussIT3MF} \ (x, \{\{[\sigma, m]\}, \lambda, \ell\}) \qquad (7.5)$$

The vertical cuts $\mathbb{A}_{(x)}(u)$ form the $FOU(\mathbb{A})$, these are IT2 FS with Gaussian IT2 MF, $\mu_{\mathbb{A}_{(x)}}(u)$ with parameters $[\sigma_u, m(x)]$ for the UMF, and for LMF: λ and ℓ. This IT3 MF is given by

$$\overline{u}(x) = exp\left[-\frac{1}{2}\left(\frac{x - m}{\sigma}\right)^2\right] \qquad (7.6)$$

$$\underline{u}(x) = \lambda \cdot exp\left[-\frac{1}{2}\left(\frac{x - m}{\sigma^*}\right)^2\right] \qquad (7.7)$$

where $\sigma^* \doteq \sigma\sqrt{\frac{\ln(\ell)}{\ln(\varepsilon)}}$, ε is an epsilon. If $\ell = 0$, $\sigma^* = \sigma$. Then $\overline{u}(x)$ and $\underline{u}(x)$ are the DOU upper and lower. The range, $\delta(u)$ and radius, σ_u of the FOU are:

$$\delta(u) = \overline{u}(x) - \underline{u}(x) \qquad (7.8)$$

$$\sigma_u = \frac{\delta(u)}{2\sqrt{3}} + \varepsilon \tag{7.9}$$

The mean, $m(x)$, of the IT3 MF $\tilde{\mu}(x, u)$, is expressed by:

$$m(x) = exp\left[-\frac{1}{2}\left(\frac{x-m}{\rho}\right)^2\right] \tag{7.10}$$

where $\rho = (\sigma + \sigma^*)/2$. Then, the vertical cuts, $\mu_{\mathbb{A}(x)}(u) = [\underline{\mu}_{\mathbb{A}(x)}(u), \overline{\mu}_{\mathbb{A}(x)}(u)]$, are described for the Equations:

$$\overline{\mu}_{\mathbb{A}(x)}(u) = exp\left[-\frac{1}{2}\left(\frac{u-u(x)}{\sigma_u}\right)^2\right] \tag{7.11}$$

$$\underline{\mu}_{\mathbb{A}(x)}(u) = \lambda \cdot exp\left[-\frac{1}{2}\left(\frac{u-u(x)}{\sigma_u^*}\right)^2\right] \tag{7.12}$$

where $\sigma_u^* = \sigma_u\sqrt{\frac{\ln(\ell)}{\ln(\varepsilon)}}$. If $\ell = 0$, then $\sigma_u^* = \sigma_u$. Then, $\overline{\mu}_{\mathbb{A}(x)}(u)$ and $\underline{\mu}_{\mathbb{A}(x)}(u)$ are the UMF and LMF of the vertical cuts IT2 FS of the secondary IT2 MF of the IT3 FS [34]. A view of this IT3 MF is depicted in Fig. 7.4.

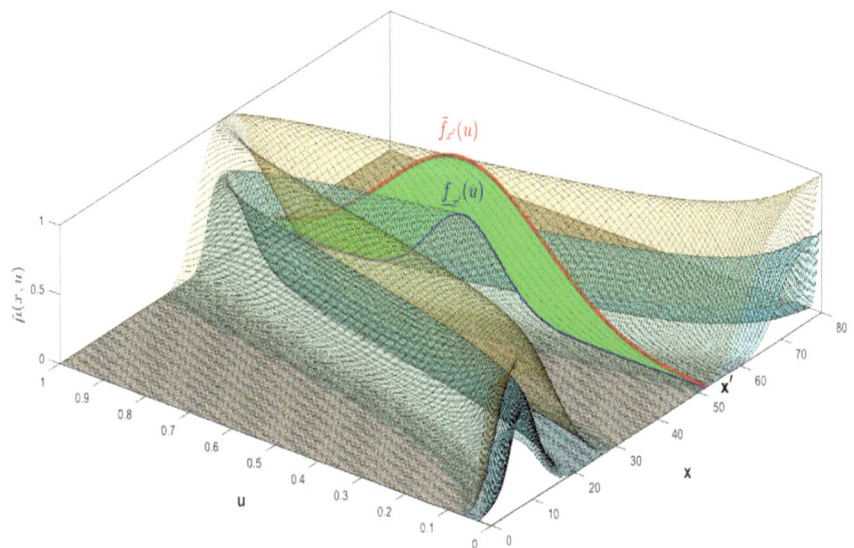

Fig. 7.4 Gaussian type-3 MF

7.2.3   Firefly Algorithm

The firefly algorithm (FA) was proposed by Xin-She Yang in [35]. FA is based on 3 principles: (1) fireflies are unisex, (2) firefly attractiveness defines its brightness, and (3) firefly brightness is postulated by the fitness function. In [36], the variation of attractiveness β using the distance r is proposed, computed by:

$$\beta = \beta_0 e^{-\gamma r^2} \tag{7.13}$$

The attractiveness at $r = 0$ is determined by β_0. The displacement of a firefly i to the brighter one j is:

$$x_i^{t+1} = x_i^t + \beta_0 e^{-\gamma r_{ij}^2}\left(x_j^t - x_i^t\right) + \alpha_t \in_i^t \tag{7.14}$$

where x_i denotes a firefly i in iteration t. Firefly attractiveness is represented with $\beta_0 e^{-\gamma r_{ij}^2}\left(x_j^t - x_i^t\right)$, and $\in_i^t$ is a vector. This vector α_t is a randomness parameter. The initial randomness scaling factor is:

$$\alpha_t = \alpha_t \delta^t \tag{7.15}$$

where δ is a value [0,1].

7.3   Method

The method and dataset are detailed in this section.

7.3.1   Method Description

The method designs ENNs, where each NN offers a prediction. The predictions are integrated employing a Type-3 FIS to estimate a weight to each prediction and compute a final prediction. FA is applied for architecture design of the ENNs and the FIS. The method is illustrated in Fig. 7.5. ENN design consists in finding the number of modules, this task is undertaken by the FA, searching from 1 to "m" modules. The FA also optimizes the FIS.

7.3.1.1   Description of the ENN

For the ENN establishment, three kinds of ANN, including function fitting [37], feedforward [38, 39], and cascade-forward [40, 41] neural network can be chosen.

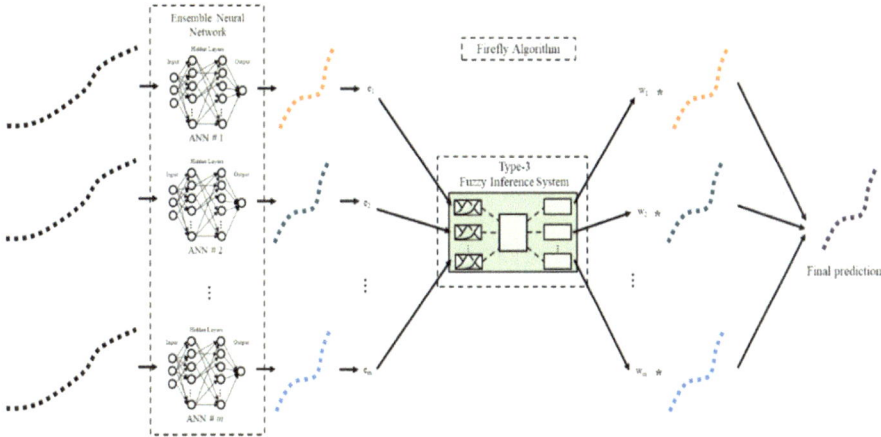

Fig. 7.5 The proposed method

Each ENN is designed using up 1 to "m" NNs, which is obtained using the FA. Using the learning phase a backpropagation algorithm was widely applied to times series prediction: Levenberg–Marquardt (LM) algorithm, with three feedback delays [15, 42]. The error of module k is:

$$MSE_k = \frac{1}{N} \sum_{i=1}^{N} (y_i - \hat{y}_{ki})^2 \qquad (7.16)$$

where the real value at time i is denoted by y_i. The prediction produced by the NN_k is denoted by $\hat{y}_{ki}$. The number of data points is given by N.

7.3.1.2 Type-3 FWA Integration

The FIS design is organized into two phases: the first one is done according to the number of NNs and establishes the FIS inputs and outputs. The ranges of the parameters, sigma, and mean of 3 Gaussian Type-3 MFs for the values ("low (L)", "medium (M)", and "high (H)") depend on the MSE. The second one (FIS design) is done by the FA, which finds values of λ, and ℓ of the MFs. In Fig. 7.6, a sample Sugeno Type-3 FIS Model is depicted.

The input ranges depend on the MSE of each module. The minimal (R_{min}) and maximal (R_{max}) values are offered by Eqs. 7.17 and 7.18 defining the range of the inputs.

$$R_{min} = \min(MSE_1, MSE_2, MSE_3, \ldots, MSE_m) \qquad (7.17)$$

$$R_{max} = \max(MSE_1, MSE_2, MSE_3, \ldots, MSE_m) \qquad (7.18)$$

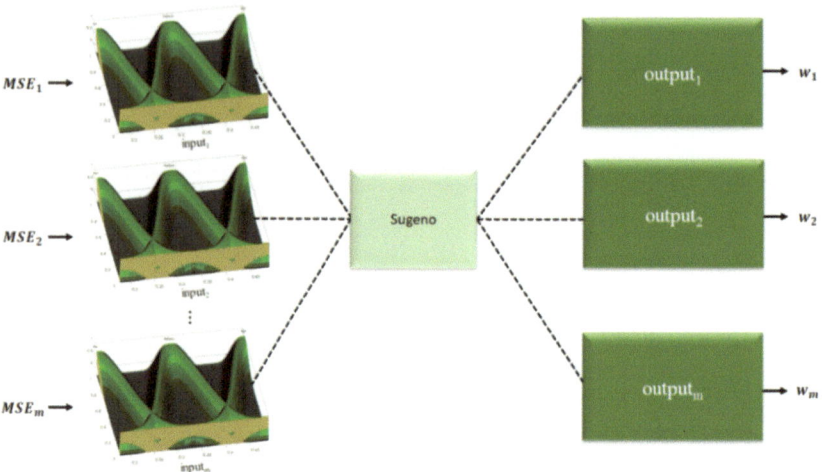

Fig. 7.6 Sugeno type-3 FIS

A case of an input variable is depicted in Fig. 7.7. A Gaussian Type-3 MF has 4 values: σ, m, λ, and ℓ. The difference between R_{min} and R_{max} is computed by Eq. 7.19 to find the σ and m values. The sigma and mean values for the 3 MFs are postulated by Eqs. 7.20–7.23. The constants of the outputs are obtained by the FA.

$$R_{dif} = R_{max} - R_{min} \tag{7.19}$$

$$\sigma = R_{dif} * 0.2 \tag{7.20}$$

Fig. 7.7 Example of a type-3 input

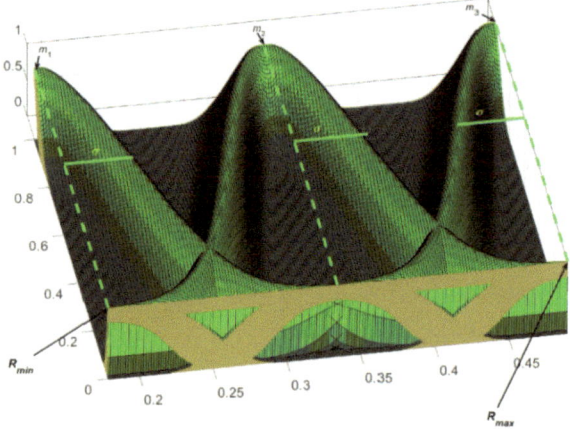

$$m_1 = R_{min} \tag{7.21}$$

$$m_2 = \left(\frac{R_{dif}}{2}\right) \tag{7.22}$$

$$m_3 = R_{max} \tag{7.23}$$

Another relevant component of the FIS is the fuzzy rules, which allow mixing predictions. The number of rules is computed with Eq. 7.24 (each variable has 3 MFs).

$$FR = 3^m \tag{7.24}$$

where m is the number of inputs. A sample ENN with 3 modules ($m = 3$) is found in Table 7.1.

The weights are utilized to calculate a final prediction using the expression:

$$P = \frac{w_1\hat{y}_1 + w_2\hat{y}_2 + \cdots + w_m\hat{y}_m}{w_1 + w_2 + \cdots + w_m} \tag{7.25}$$

where w_1 is the prediction weight of ANN #1 and so on up to w_m, which is the prediction weight obtained by ANN m, $\hat{y}_1$ is the prediction of ANN #1, and so on up to $\hat{y}_m$, which is the prediction of ANN m.

The FA search space is defined in Table 7.2. The search space for the ensemble neural network is established based on previous works applied to time series prediction [11, 15].

For each experiment, the firefly algorithm is established with the parameters based on [15, 43]. The FA employs 10 fireflies, α of 0.01, β of 1, δ of 0.97, and 30 iterations. In Fig. 7.8, the optimization flowchart is depicted. The FA objective is the minimization of final ENN prediction error and the objective function is postulated as:

$$f = \frac{1}{N}\sum_{i=1}^{N}(Y_i - P_i)^2 \tag{7.26}$$

where the real value at time i is Y_i, the final ENN prediction is P_i, and the number of points is denoted by N.

Table 7.1 Fuzzy rules for the inputs and outputs

	Antecedents			Consequents		
Rule	e_1	e_2	e_3	w_1	w_2	w_3
1	L	L	L	H	H	H
2	L	L	M	H	H	M
3	L	L	H	H	H	L
4	L	M	L	H	M	H
5	L	M	M	H	M	M
6	L	M	H	H	M	L
7	L	H	L	H	L	H
8	L	H	M	H	L	M
9	L	H	H	H	L	L
10	M	L	L	M	H	H
11	M	L	M	M	H	M
12	M	L	H	M	H	L
13	M	M	L	M	M	H
14	M	M	M	M	M	M
15	M	M	H	M	M	L
16	M	H	L	M	L	H
17	M	H	M	M	L	M
18	M	H	H	M	L	L
19	H	L	L	L	H	H
20	H	L	M	L	H	M
21	H	L	H	L	H	L
22	H	M	L	L	M	H
23	H	M	M	L	M	M
24	H	M	H	L	M	L
25	H	H	L	L	L	H
26	H	H	M	L	L	M
27	H	H	H	L	L	L

Table 7.2 Search space

	Parameters	Minimum	Maximum
ENN	m	2	5
	Hidden Layers (h)	1	5
	Neurons	1	50
	Goal Error	0.00001	0.001
Type-3 FIS	λ	0.10	0.90
	ℓ	0.10	0.90
	Constants	0.10	0.90

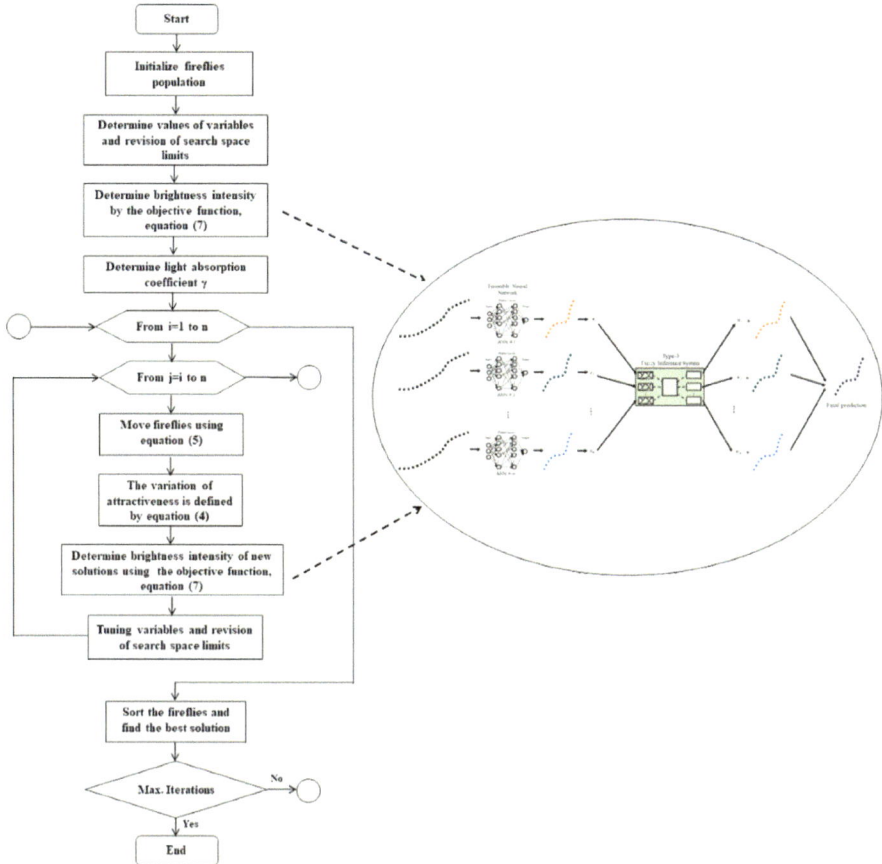

Fig. 7.8 Flowchart of the FA

7.3.2   Dataset Description

The information of the confirmed cases dataset worldwide is from the Humanitarian Data Exchange [44]. The data period is from 01/22/20 to 03/29/22, which is 798 days. For the data period, 12 countries are analyzed. In Fig. 7.9, the information by the country is shown.

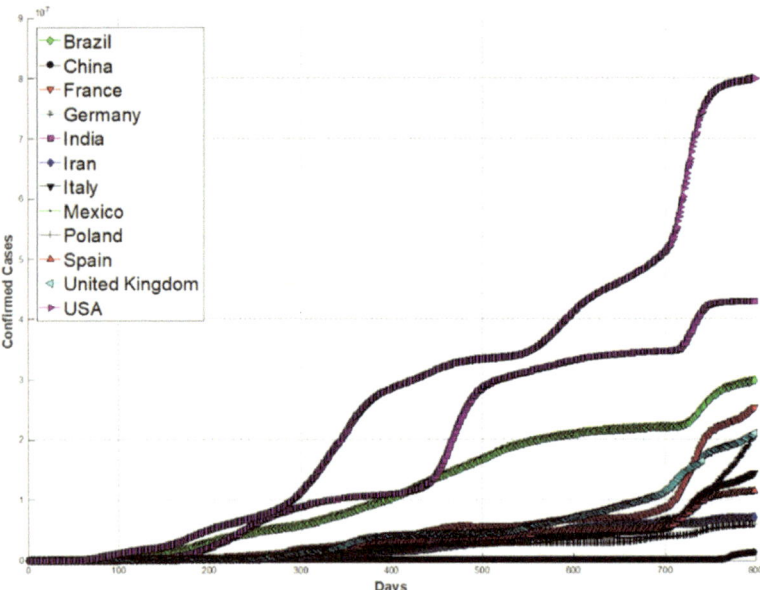

Fig. 7.9 Data in the period

7.4 Experimental Results

The experimentation was done with a testing set of 30% of the data (black points), leaving 70% to be split into 2 sets: training and validation (80/20). Comparisons versus a Type-2 FWA presented by [15] are performed. Confirmed cases of 12 countries are employed, performing 30 runs per country. Prediction of 20 future days is estimated by the ENN (pink points). The predictions are aggregated utilizing the weights offered by the FIS, and finally, the prediction is computed by Eq. 7.25.

Type-3 results are compared with Type-2. For the testing set, the results achieved (MSE) are illustrated in Table 7.3.

However, for the future days, in general (except for Brazil), Type-3 FWA obtained better performance as summarized in Table 7.4 (bold indicates best results).

For the testing data and the next 20 days, the achieved results are graphical illustrated in Figs. 7.10 and 7.11.

7.5 Conclusions

This chapter utilized COVID-19 cases information for prediction. That means that both integration techniques provided almost the same results. For this work, the evaluation of the future days is the most important part of the evaluation, demonstrating

Table 7.3 Results for the period (Testing prediction)

Country	FWA					
	Type-2			Type-3		
	Best	Average	Worst	Best	Average	Worst
Brazil	0.000000688	0.000000709	0.000000738	0.000000690	**0.00000070**	0.000000790
China	0.000013600	0.000014200	0.000015400	0.000013600	**0.00001410**	0.000015200
France	0.000006090	**0.000006250**	0.000006800	0.000006140	0.00000636	0.000006810
Germany	0.000001560	**0.000001940**	0.000002300	0.000001610	0.00000202	0.000002840
India	0.000000049	**0.000000050**	0.000000058	0.000000049	0.00000005	0.000000058
Iran	0.000000213	**0.000000223**	0.000000233	0.000000214	0.00000022	0.000000246
Italy	0.000002560	**0.000002580**	0.000002610	0.000002550	0.00000259	0.000002640
Mexico	0.000007270	**0.000007280**	0.000007300	0.000007270	0.00000729	0.000073200
Poland	0.000000455	**0.000000460**	0.000000498	0.000000455	0.00000046	0.000000478
Spain	0.000022000	**0.000022100**	0.000022200	0.000022000	0.00002210	0.000022300
United Kingdom	0.000008420	0.000008550	0.000008800	0.000008360	**0.00000854**	0.000008790
USA	0.000002860	**0.000002870**	0.000002890	0.000002850	0.00000288	0.000002920

Table 7.4 Results for the period (Future days)

Country	FWA					
	Type-2			Type-3		
	Best	Avg	Worst	Best	Avg	Worst
Brazil	0.00000304	**0.01970**	0.1420	0.000001840	0.0246	0.1060
China	0.00184000	0.06980	0.2960	0.000522000	**0.0297**	0.1610
France	0.00000607	0.02060	0.1940	0.000004130	**0.0071**	0.0606
Germany	0.00080200	0.08550	0.4110	0.000040900	**0.0302**	0.1010
India	0.00000015	0.00889	0.1540	0.000000305	**0.0030**	0.0205
Iran	0.00001120	0.01780	0.0982	0.000000756	**0.0143**	0.1050
Italy	0.00000757	0.04540	0.2920	0.000012400	**0.0176**	0.0832
Mexico	0.00002860	0.00914	0.1810	0.000014800	**0.00149**	0.0300
Poland	0.00008310	0.02050	0.4280	0.000057100	**0.00608**	0.0585
Spain	0.00058200	0.000917	0.0015	0.000556000	**0.00083**	0.0015
United Kingdom	0.00002800	0.00707	0.1620	0.000269000	**0.01260**	0.0998
USA	0.00000315	0.00802	0.0604	0.000000626	**0.00532**	0.0949

the effectiveness or not of a method. For the future days, for the second data period (158 days), there is no significant difference in the comparison between both integration techniques, which means that both integration techniques provide similar results. However, for the data period (798 days), the proposed method using Type-3 FWA provides statistically better results than Type-2. With the achieved results,

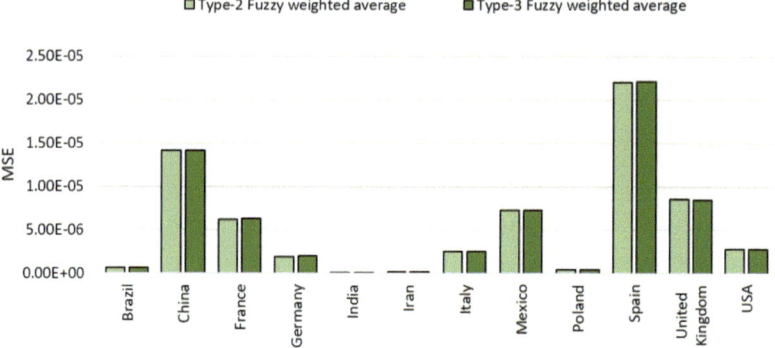

Fig. 7.10 Data period (testing)

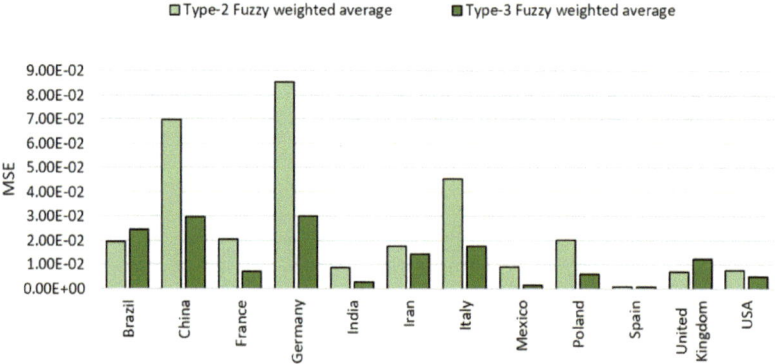

Fig. 7.11 Data period (future prediction)

we can state that Type-3 FWA offers better results in prediction. In future works, another data period with more information and other optimization methods will be implemented to compare results, like for example the ones in [45–50]. We could also consider other applications for the proposed approach, as the ones in [51–57].

References

1. Z. Jin, J.-Y. Liu, R. Feng, L. Ji, Z.-L. Jin, H.-B. Li, Drug treatment of coronavirus disease 2019 (COVID-19) in China. Eur. J. Pharmacol.Pharmacol. **883**, 1–7 (2020)
2. Q. Zhang, Y. Wei, M. Chen, Q. Wan, X. Chen, Clinical analysis of risk factors for severe COVID-19 patients with type 2 diabetes. J. Diabetes Complicat. **34**(10), 1–5 (2020)
3. P. Melin, J. Monica, D. Sánchez, O. Castillo, Analysis of spatial spread relationships of coronavirus (COVID-19) pandemic in the world using self organizing maps. Chaos Solitons Fractals **138**, 1–7 (2020)

4. D. Reddy, V. Atam, P. Rai, F. Khan, S. Pandey, H. Malhotra, K. Gupta, S. Sonkar, R. Verma, COVID-19 cases and their outcome among patients with uncommon co-existing illnesses: a lesson from Northern India. Clin. Epidemiol. Health **15**, 1–6 (2022)
5. L. Zha, T. Sobue, A. Hirayama, T. Takeuchi, K. Tanaka, Y. Katayama, S. Komukai, T. Shimazu, T. Kitamura, COVID-19 epidemiology research group, "Characteristics and outcomes of COVID-19 in reproductive-aged pregnant and nonpregnant women in Osaka, Japan." Int. J. Infect. Dis. **117**, 195–200 (2022)
6. L. Reyes, A. Rodriguez, A. Bastidas, D. Parra-Tanoux, Y.V. Fuentes, E. García-Gallo, G. Moreno, G. Ospina-Tascon, G. Hernandez, E. Silva, A.M. Díaz, M. Jibaja, M. Vera, E. Díaz, M. Bodí, J. Solé-Violán, R. Ferrer, A. Albaya-Moreno, L. Socias, Á. Estella, A. Loza-Vazquez, R. Jorge-García, I. Sancho, I. Martin-Loeches, Dexamethasone as risk-factor for ICU-acquired respiratory tract infections in severe COVID-19. J. Crit. Care **69**, 1–8 (2022)
7. D. Liu, W. Ding, Z. Dong, W. Pedrycz, Optimizing deep neural networks to predict the effect of social distancing on COVID-19 spread. Comput. Ind. Eng.. Ind. Eng. **166**, 1–17 (2022)
8. Y. Kuvvetli, M. Deveci, T. Paksoy, H. Garg, A predictive analytics model for COVID-19 pandemic using artificial neural networks. Decis. Anal. J. **1**, 1–13 (2021)
9. H. Verma, S. Mandal, A. Gupta, Temporal deep learning architecture for prediction of COVID-19 cases in India. Expert Syst. Appl. **195**, 1–11 (2022)
10. S. Khalilpourazari, H. Doulabi, A. Çiftçioglu, G. Weber, Gradient-based grey wolf optimizer with Gaussian walk: application in modelling and prediction of the COVID-19 pandemic. Expert Syst. Appl. **177**, 1–23 (2021)
11. M. Pulido, P. Melin, O. Castillo, Particle swarm optimization of ensemble neural networks with fuzzy aggregation for time series prediction of the Mexican Stock Exchange. Inf. Sci. **280**, 188–204 (2014)
12. P. Melin, J. Monica, D. Sánchez, O. Castillo, Multiple ensemble neural network models with fuzzy response aggregation for predicting COVID-19 time series: the case of Mexico. Healthcare **8**(2), 1–13 (2020)
13. D. Jia, Z. Wu, Seismic fragility analysis of RC frame-shear wall structure under multidimensional performance limit state based on ensemble neural network. Eng. Struct.Struct. **246**, 1–15 (2021)
14. I. Wilkinson, R. Bhattacharjee, J. Shafer, A. Osborne, Confidence estimation in the prediction of epithermal neutron resonance self-shielding factors in irradiation samples using an ensemble neural network. Energy AI **7**, 1–19 (2022)
15. P. Melin, D. Sánchez, J. Monica, O. Castillo, Optimization using the firefly algorithm of ensemble neural networks with type-2 fuzzy integration for COVID-19 time series prediction. Soft. Comput.Comput. **1**, 1–38 (2021)
16. Z. Liu, A. Mohammadzadeh, H. Turabieh, M. Mafarja, S. Band, A. Mosavi, A new online learned interval type-3 fuzzy control system for solar energy management systems. IEEE Access **9**, 10498–10508 (2021)
17. S. Qasem, A. Ahmadian, A. Mohammadzadeh, S. Rathinasamy, B. Pahlevanzadeh, A type-3 logic fuzzy system: optimized by a correntropy based Kalman filter with adaptive fuzzy kernel size. Inf. Sci. **572**, 424–443 (2021)
18. Y. Cao, A. Raise, A. Mohammadzadeh, S. Rathinasamy, S. Band, A. Mosavi, Deep learned recurrent type-3 fuzzy system: application for renewable energy modeling/prediction. Energy Rep. **7**, 8115–8127 (2021)
19. S. Hanandeh, Introducing mathematical modeling to estimate pavement quality index of flexible pavements based on genetic algorithm and artificial neural networks. Case Stud. Constr. Mater. **16**, 1–13 (2022)
20. V. Tam, A. Butera, K. Le, L. Da Silva, A. Evangelista, A prediction model for compressive strength of CO_2 concrete using regression analysis and artificial neural networks. Constr. Build. Mater. **324**, 1–13 (2022)
21. C. Aggarwal, *Neural Networks and Deep Learning: A Textbook*, 1st edn. (Springer, 2018)
22. B. Peng, L. Tong, D. Yan, W. Huo, Experimental research and artificial neural network prediction of free piston expander-linear generator. Energy Rep. **8**, 1966–1978 (2022)

23. K. Prakarsha, G. Sharma, Time series signal forecasting using artificial neural networks: An application on ECG signal. Biomed. Signal Process. Control **76**, 1–10 (2022)
24. K. Gurney, *An Introduction to Neural Networks*, 1st edn. (CRC Press, 1997)
25. S. Haykin, *Neural Networks: A Comprehensive Foundation*, 2nd edn. (Prentice Hall, 1998).
26. L. Zadeh, Fuzzy sets. Inf. Control. **8**(3), 338–353 (1965)
27. L. Zadeh, The concept of a linguistic variable and its application to approximate reasoning. Inf. Sci. **8**(3), 199–249 (1975)
28. L. Zadeh, Some reflections on soft computing, granular computing and their roles in the conception, design and utilization of information/intelligent systems. Soft. Comput.Comput. **2**, 23–25 (1998)
29. P. Melin, O. Castillo, *Hybrid Intelligent Systems for Pattern Recognition Using Soft Computing: An Evolutionary Approach for Neural Networks and Fuzzy Systems*, 1st edn. (Springer, 2005)
30. H. Al-Jamimi, T. Saleh, Transparent predictive modelling of catalytic hydrodesulfurization using an interval type-2 fuzzy logic. J. Clean. Prod. **231**, 1079–1088 (2019)
31. P. Melin, O. Castillo, A review on type-2 fuzzy logic applications in clustering, classification and pattern recognition. Appl. Soft Comput.Comput. **21**, 568–577 (2014)
32. J. Rickard, J. Aisbett, G. Gibbon, Fuzzy subsethood for fuzzy sets of type-2 and generalized type-n. IEEE Trans. Fuzzy Syst. **17**(1), 50–60 (2009)
33. A. Mohammadzadeh, M. Sabzalian, W. Zhang, An interval type-3 fuzzy system and a new online fractional-order learning algorithm: theory and practice. IEEE Trans. Fuzzy Syst. **28**(9), 1940–1950 (2020)
34. O. Castillo, J. Castro, P. Melin, *Interval Type-3 Fuzzy Systems: Theory and Design* (Springer, 2022)
35. X. Yang, Firefly algorithms for multimodal optimization, in *Proceeding 5th Symposium on Stochastic Algorithms, Foundations and Applications*, vol. 5792, (2009), pp. 169–178
36. X. Yang, X. He, Firefly algorithm: recent advances and applications. Int. J. Swarm Intell. **1**(1), 36–50 (2013)
37. Z. Chen, A. Ashkezari, I. Tlili, Applying artificial neural network and curve fitting method to predict the viscosity of SAE50/MWCNTs-TiO2 hybrid nanolubricant. Phys. A Stat. Mech. Appl. **549**, 1–11 (2020)
38. Z.-G. Che, T.-A. Chiang, Z.-H. Che, Feed-forward neural networks training: a comparison between genetic algorithm and back-propagation learning algorithm. Int. J. Innov. Comput. Inf. Control **7**(10), 5839–5850 (2011)
39. J. Gauthier, P. Micheau, Feedfoward and feedback adaptive controls for continuously variable transmissions. IFAC Proc. Vol. **45**(16), 1460–1465 (2012)
40. Y. An, K. Yoo, M. Na, Y.-S. Kim, Critical flow prediction using simplified cascade fuzzy neural networks. Ann. Nucl. Energy **136**, 1–11 (2020)
41. Ü. Budak, Y. Guo, E. Tanyildizi, A. Şengür, Cascaded deep convolutional encoder-decoder neural networks for efficient liver tumor segmentation. Med. Hypotheses **134**, 1–8 (2020)
42. M. Pulido, P. Melin, Optimization of ensemble neural networks with type-2 fuzzy integration of responses for the Dow Jones time series prediction. Intell. Autom. Soft Comput. **20**, 403–418 (2014)
43. D. Sánchez, P. Melin, O. Castillo, Optimization of modular granular neural networks using a firefly algorithm for human recognition. Eng. Appl. Artif. Intell.Intell. **64**, 172–186 (2017)
44. The Humanitarian Data Exchange (HDX). (2022, April). https://data.humdata.org/dataset/novel-coronavirus-2019-ncov-cases
45. F. Valdez, P. Melin, O. Castillo, Evolutionary method combining particle swarm optimization and genetic algorithms using fuzzy logic for decision making, in *IEEE International Conference on Fuzzy Systems*, (2009), pp. 2114–2119
46. F. Valdez, J.C. Vazquez, P. Melin, O. Castillo, Comparative study of the use of fuzzy logic in improving particle swarm optimization variants for mathematical functions using co-evolution. Appl. Soft Comput.Comput. **52**, 1070–1083 (2017)
47. O. Castillo, E. Lizarraga, J. Soria, P. Melin, F. Valdez, New approach using ant colony optimization with ant set partition for fuzzy control design applied to the ball and beam system. Inf. Sci. **294**, 203–215 (2015)

48. L. Amador-Angulo, O. Mendoza, J.R. Castro, A. Rodriguez-Diaz, P. Melin, O. Castillo, Fuzzy sets in dynamic adaptation of parameters of a bee colony optimization for controlling the trajectory of an autonomous mobile robot. Sensors **16**(9), 1458 (2016)

49. F. Valdez, H. Carreon-Ortiz, O. Castillo, CMOA—Continuous Mycorrhiza Optimization Algorithm, in *Mycorrhiza Optimization Algorithm.* SpringerBriefs in Applied Sciences and Technology. (Springer, Cham, 2023). https://doi.org/10.1007/978-3-031-47369-2_5

50. F. Valdez, H. Carreon-Ortiz, O. Castillo, DMOA—Discrete Mycorrhiza Optimization Algorithm, in *Mycorrhiza Optimization Algorithm.* SpringerBriefs in Applied Sciences and Technology. (Springer, Cham, 2023). https://doi.org/10.1007/978-3-031-47369-2_6

51. E. Ontiveros, P. Melin, O. Castillo, Comparative study of interval type-2 and general type-2 fuzzy systems in medical diagnosis. Inf. Sci. **525**, 37–53 (2020)

52. J.R. Castro, O. Castillo, P. Melin, A. Rodriguez-Diaz, Building fuzzy inference systems with a new interval type-2 fuzzy logic toolbox. Trans. Comput. Sci. I, 104–114 (2008).

53. D. Sanchez, P. Melin, O. Castillo, A grey wolf optimizer for modular granular neural networks for human recognition. Comput. Intell. Neurosci. **2017**, 4180510:1–4180510:26 (2017)

54. O. Castillo, P. Melin, Intelligent adaptive model-based control of robotic dynamic systems with a hybrid fuzzy-neural approach. Appl. Soft Comput.Comput. **3**(4), 363–378 (2003)

55. M.H.F. Zarandi, A.A.S. Asl, S. Sotudian, O. Castillo, A state of the art review of intelligent scheduling. Artif. Intell. Rev.. Intell. Rev. **53**, 501–593 (2020)

56. H.I. Seker, S. Kacar, O. Castillo, S. Uzun, I. Pehlivan, Z. Tatli, Detection of resistance spot welding faults in copper materials by transfer learning method. Appl. Comput. Math. **22**(3), 430–445 (2023). https://doi.org/10.30546/1683-6154.22.3.2023.430

57. F. Valdez, O. Castillo, P. Cortes-Antonio, P. Melin, Applications of intelligent optimization algorithms and fuzzy logic systems in aerospace: a review. Appl. Comput. Math. **21**(3), 233–245 (2022). https://doi.org/10.30546/1683-6154.21.3.2022.233

Chapter 8
Conclusions of Type-3 Fuzzy Logic in Prediction

In this chapter the basic terminology and methods for designing type-3 fuzzy sets, membership functions, inference and fuzzy systems for prediction have been outlined in this book [1]. Type-2 fuzzy system design for prediction is a challenging endeavor. So, for the case of designing type-3 fuzzy predictors, the problem is more complicated due to the higher number of parameters to consider, and we have involved the utilization of nature-inspired optimization techniques for this problem [2].

The main idea of the work has been to introduce in a systematic way the concepts and methods of interval type-3 fuzzy systems that with their higher uncertainty handling capabilities we expect that will solve, in a better way, more difficult prediction problems [3, 4].

This monograph offers type-3 fuzzy applications to prediction and forecasting. In Chap. 3 type-3 fuzzy logic in prediction is highlighted. We also presented in Chap. 4 a general approach for Prediction of COVID-19 with Type-3 and Fractal Theory. In Chap. 5 the type-3 fuzzy aggregation of NNs for prediction approach is outlined. In Chap. 6 an approach for type-3 aggregators for ENNs in prediction was presented. In Chap. 7 the optimal design of type-3 fuzzy systems and ENNs using FA was outlined. Finally, in Chapter 8 some conclusions are outlined.

As future works, we consider that it will be interesting to build type-3 fuzzy systems for specific problems in different areas of application, such as intelligent control [5, 6], robotics [7, 8], pattern recognition [9], optimization [10, 11], diagnosis [12], and other ones [13–15].

There are also nature-inspired algorithms that have not yet been applied in designing type-3, like for example: plant self-defense algorithm, electromagnetism-based algorithm, and similar ones [16, 17]. In addition, another research area has been the utilization of fuzzy systems for achieving adaptation in metaheuristics, and we envision a direct utilization of type-3 in this area, as is presented in [18–20].

Finally, it is worthwhile to mention other possible avenues of research, which would be the hybrid combination of T3FL with Z numbers, also a possible hybrid type-3 neutrosophic approach could be envisioned, as well as considering type-3

© The Author(s), under exclusive license to Springer Nature Switzerland AG 2024   95
O. Castillo and P. Melin, *Type-3 Fuzzy Logic in Time Series Prediction*,
SpringerBriefs in Computational Intelligence,
https://doi.org/10.1007/978-3-031-59714-5_8

Pitagorean fuzzy systems or a type-3 hesitant combination could be think of, and so on other similar interesting proposals [21].

References

1. O. Castillo, J.R. Castro, P. Melin, *Interval Type-3 Fuzzy Systems: Theory and Design* (Springer, Cham, Switzerland, 2022)
2. O. Castillo, P. Melin, Towards interval type-3 intuitionistic fuzzy sets and systems. Mathematics **10**, 4091 (2022). https://doi.org/10.3390/math10214091
3. Z. Liu, A. Mohammadzadeh, H. Turabieh, M. Mafarja, S.S. Band, A. Mosavi, A new online learned interval type-3 fuzzy control system for solar energy management systems. IEEE Access **9**, 10498–10508 (2021)
4. R.H. Vafaie, A. Mohammadzadeh, M. Piran, A new type-3 fuzzy predictive controller for MEMS gyroscopes. Nonlinear Dyn.Dyn. **106**(1), 381–403 (2021)
5. L. Aguilar, P. Melin, O. Castillo, Intelligent control of a stepping motor drive using a hybrid neuro-fuzzy ANFIS approach. Appl. Soft Comput. **3**(3), 209–219
6. P. Melin, O. Castillo, Adaptive intelligent control of aircraft systems with a hybrid approach combining neural networks, fuzzy logic and fractal theory. Appl. Soft Comput. **3**(4), 353–362 (2003)
7. O. Castillo, P. Melin, A new fuzzy-fractal-genetic method for automated mathematical modelling and simulation of robotic dynamic systems, in *1998 IEEE International Conference on Fuzzy Systems (FUZZ-IEEE 1998) Proceedings*, vol. 2, pp. 1182–1187
8. O. Castillo, P. Melin, Intelligent adaptive model-based control of robotic dynamic systems with a hybrid fuzzy-neural approach. Appl. Soft Comput. **3**(4), 363–378 (2003)
9. D. Sanchez, P. Melin, O. Castillo, A grey wolf optimizer for modular granular neural networks for human recognition. Comput. Intell. Neurosci. **2017**, 4180510:1–4180510:26 (2017)
10. F. Valdez, P. Melin, O. Castillo, Evolutionary method combining particle swarm optimization and genetic algorithms using fuzzy logic for decision making, in *IEEE International Conference on Fuzzy Systems* (2009), pp. 2114–2119
11. F. Valdez, J.C. Vazquez, P. Melin, O. Castillo, Comparative study of the use of fuzzy logic in improving particle swarm optimization variants for mathematical functions using co-evolution. Appl. Soft Comput. **52**, 1070–1083 (2017)
12. E. Ontiveros, P. Melin, O. Castillo, Comparative study of interval type-2 and general type-2 fuzzy systems in medical diagnosis. Inf. Sci. **525**, 37–53 (2020)
13. H.I. Seker, S. Kacar, O. Castillo, S. Uzun, I. Pehlivan, Z. Tatli, Detection of resistance spot welding faults in copper materials by transfer learning method. Appl. Comput. Math. **22**(3), 430–445 (2023). https://doi.org/10.30546/1683-6154.22.3.2023.430
14. D. Mohapatra, S. Chakraverty, O. Castillo, Numerical investigation of fluid dynamic model in uncertain environment. Appl. Comput. Math. **22**(3), 297–316 (2023). https://doi.org/10.30546/1683-6154.22.3.2023.297
15. F. Valdez, O. Castillo, P. Cortes-Antonio, P. Melin, Applications of intelligent optimization algorithms and fuzzy logic systems in aerospace: a review. Appl. Comput. Math. **21**(3), 233–245 (2022). https://doi.org/10.30546/1683-6154.21.3.2022.233
16. F. Valdez, H. Carreon-Ortiz, O. Castillo, CMOA—Continuous Mycorrhiza Optimization Algorithm, in *Mycorrhiza Optimization Algorithm.* SpringerBriefs in Applied Sciences and Technology (Springer, Cham, 2023). https://doi.org/10.1007/978-3-031-47369-2_5
17. F. Valdez, H. Carreon-Ortiz, O. Castillo, DMOA—Discrete Mycorrhiza Optimization Algorithm, in *Mycorrhiza Optimization Algorithm.* SpringerBriefs in Applied Sciences and Technology (Springer, Cham, 2023). https://doi.org/10.1007/978-3-031-47369-2_6
18. L. Amador-Angulo, O. Castillo, P. Melin, J.R. Castro, Interval type-3 fuzzy adaptation of the bee colony optimization algorithm for optimal fuzzy control of an autonomous mobile robot. Micromachines **13**, 1490 (2022). https://doi.org/10.3390/mi13091490

19. L. Amador-Angulo, O. Castillo, J.R. Castro et al., A new approach for interval type-3 fuzzy control of nonlinear plants. Int. J. Fuzzy Syst. **25**, 1624–1642 (2023). https://doi.org/10.1007/s40815-023-01470-9

20. C. Peraza, P. Ochoa, O. Castillo, Z.W. Geem, Interval-type 3 fuzzy differential evolution for designing an interval-type 3 fuzzy controller of a unicycle mobile robot. Mathematics **10**, 3533 (2022). https://doi.org/10.3390/math10193533

21. O. Castillo, P. Melin, Proposal for mediative fuzzy control: from type-1 to type-3. Symmetry **2023**, 15 (1941). https://doi.org/10.3390/sym15101941

Index